T0318239

Innovation Research in Technology and Engineering Management

Philosophy may not seem to be an obvious source to discover methods for successful product innovation management. However, this book shows that systematic reflection on the nature of product innovation management, supported by insights from the philosophy of technology, can illuminate the innovation process in technology and engineering.

Presenting methodological guidelines and philosophical reflections, this book guides readers through each phase of product innovation. At each step, ideas from the philosophy of technology are translated into practical guidelines for managing these processes. The book works through the philosophical perspectives on innovation, methods in innovation design and research, and the value and ethical implications of innovation.

Bridging the gap between philosophical context and practical methodologies, this book will be highly valuable for postgraduate students and academics researching and teaching innovation and philosophy of technology.

Marc J. de Vries is Professor of Christian Philosophy of Technology and Professor of Science Education, both at Delft University of Technology, the Netherlands. He is also the coauthor of *Philosophy of Technology: An Introduction for Technology and Business Students* (Routledge, 2016).

Routledge Focus on Business and Management

The fields of business and management have grown exponentially as areas of research and education. This growth presents challenges for readers trying to keep up with the latest important insights. *Routledge Focus on Business and Management* presents small books on big topics and how they intersect with the world of business research.

Individually, each title in the series provides coverage of a key academic topic, whilst collectively, the series forms a comprehensive collection across the business disciplines.

Human Resource Management in the Indian Tea Industry
Nirmal Chandra Roy and Debasish Biswas

Entrepreneurship for Rural Start-ups
Lessons and Guidance for New Venture Creation
Edited by Gloria Jiménez-Marín, Alejandro López Rodríguez, Miguel Torres García and José Guadix Martín

Entrepreneurship and Work in the Gig Economy
The Case of the Western Balkans
Mirjana Radović-Marković, Borislav Đukanović, Dušan Marković and Arsen Dragojević

Innovation Research in Technology and Engineering Management
A Philosophical Approach
Marc J. de Vries

For more information about this series, please visit: www.routledge.com/Routledge-Focus-on-Business-and-Management/book-series/FBM

Innovation Research in Technology and Engineering Management

A Philosophical Approach

Marc J. de Vries

LONDON AND NEW YORK

First published 2021
by Routledge
2 Park Square, Milton Park, Abingdon, Oxon OX14 4RN

and by Routledge
605 Third Avenue, New York, NY 10158

Routledge is an imprint of the Taylor & Francis Group, an informa business

British Library Cataloguing-in-Publication Data
A catalogue record for this book is available from the British Library

Library of Congress Cataloging-in-Publication Data
Names: Vries, Marc J. de, 1958– author.
Title: Innovation research in technology and engineering management : a philosophical approach / Marc J. de Vries.
Description: New York : Routledge, 2021. |
Series: Routledge focus on business and management |
Includes bibliographical references and index.
Identifiers: LCCN 2020053677 (print) | LCCN 2020053678 (ebook) | ISBN 9780367821067 (hardback) |
ISBN 9781003013556 (ebook)
Subjects: LCSH: New products—Management. | Industrial management—Technological innovations. | Technological innovations—Moral and ethical aspects.
Classification: LCC HF5415.153 .V75 2021 (print) |
LCC HF5415.153 (ebook) | DDC 658.5/75—dc23
LC record available at https://lccn.loc.gov/2020053677
LC ebook record available at https://lccn.loc.gov/2020053678

ISBN: 978-0-367-82106-7 (hbk)
ISBN: 978-0-367-67769-5 (pbk)
ISBN: 978-1-003-01355-6 (ebk)

Typeset in Times New Roman
by codeMantra

Contents

Preface

This book was written from the background of more than 10 years of experience in supervising master students doing their thesis work on product development in industrial settings, and another 20 years of more theoretical work on the philosophy of technological developments. I did my supervision as a member of a philosophy and ethics of technology group, and therefore, the focus of the students' work was always somehow related to philosophical and methodological reflections on product development. This sounds like something almost impossible. How can philosophy as perhaps the most abstract and theoretical discipline make a practical contribution to product development in industrial corporations? The answer lies in the fact that philosophy helps to ask the right questions. Product development and its management is full of assumptions that tend to be taken for granted while in fact they may be more problematic than thought. My aim in writing this book is to make philosophy of technology fruitful for product innovation management by showing what kind of assumptions need to be given attention and questioned in the practice of that field.

In this I tried to get to the bottom of that by showing that these assumptions are always in some way ultimately related to our deepest convictions about reality. These can be convictions about the different natures of the things around us (for instance, the difference between nature and man-made objects), convictions about what we can know and the different ways in which we can know things. They can be convictions about what are clever and what are clumsy ways of performing different types of activities. They can also be convictions about what is morally acceptable and what is not. All these are matters of our fundamental ideas about the reality in which people do their product innovation management work. The coherent set of such ideas held by an individual or a group are called a 'worldview'. There is nothing esoteric about that. We all hold worldviews, and most of the time they remain hidden and implicit, but sometimes it is necessary to become aware of them so that we can prevent using assumptions that may not be fruitful.

In the beginning of the book, I show different options for worldviews people can and do hold, but in the course of the book I will be increasingly explicit about my own conviction that worldviews that strongly reduce reality (e.g., to only matter, or only money) tend to oversimplify and do not do justice to the complexity of reality. This complexity should not only be seen as problematic but, in the first place, as a richness to be exploited. In the philosophical tradition in which I do my more theoretical work, a non-reductionist worldview is used that has a background in Christian belief. My experience in teaching this approach to both Christian and non-Christian students is that both groups almost equally appreciate that when this Christian perspective is 'translated' into the practical realm of technological development, it makes a lot of sense, and does not require subscription to all of Christian belief. Non-reductionism is also 'preached' by other worldviews, and that makes the perspective of Christian philosophy easily recognizable for people holding other non-reductionist worldviews.

In the tradition of Christian philosophy as initiated by Herman Dooyeweerd, a 20th-century Dutch philosopher, aspects of the non-reductionist worldview have been elaborated into a way of conceptualizing reality and what we can know of it, and in this elaboration lies a specific contribution to product innovation management. The result is a book that certainly is not of a cookbook nature. Perhaps it is rather of an anti-cookbook nature because cookbooks are almost necessarily simplifications of the very complex nature of the real chef's work. In the years that I worked with various companies through the master's students I supervised, I have observed many a times how philosophical reflections worked as 'eye-openers' for managers and made them become more aware of the value of this discipline for practice. I hope that this publication will be helpful for practitioners to see how relevant philosophical reflections can be for product innovation management.

Delft, the Netherlands, September 2020

Marc J. de Vries

1 Introduction

"When he who hears does not know what he who speaks means, and when he who speaks does not know what he himself means – that is philosophy". This quote from Voltaire, the famous 18th-century French philosopher, does not exactly sound as an invitation to start looking in the direction of philosophy for finding guidelines in product innovation management. Here are some more examples of what of course should be read as self-mockery by philosophers. William James, American pragmatist philosopher, says: "There is only one thing that a philosopher can be relied on to do, and that is to contradict other philosophers". Here comes Montaigne, 16th-century French philosopher: "Wonder is the foundation of all philosophy, inquiry the progress, ignorance the end". Finally Jonathan Swift, who can be considered to be a (political) philosopher even though we mostly see him as an essayist, says: "The various opinions of philosophers have scattered through the world as many plagues of the mind as Pandora's box did those of the body; only with this difference, that they have not left hope at the bottom".

All these quotes have been taken from the first pages of a book that aims to introduce philosophy to a wide audience ("Philosophy for Dummies", by Tom Morris; Morris 1999). Evidently, Morris found them useful to whet the reader's appetite for studying philosophy. And of course, the humor in them helps to change our image of philosophers as if they are so abstract in orientation that they cannot have a good sense of humor. Also, they can be quite involved in practical matters such as politics and technology. As Allyn Fives argues, philosophers can have a unique contribution to public debates (Fives, 2016). In this book on technology and engineering, it will be argued that philosophy also can be a rich source of inspiration to support product innovation management and research to support that.

'Philosophy' is a word that is used in different ways. Often it means something like 'ideas' or 'beliefs'. "My philosophy about this is that . . ." is something we regularly hear or read. In that meaning it does not have much authority, depending of course on who says or writes it. After all, my ideas can be based on flawed assumptions or being ill-informed. Philosophy, however, is more than just ideas I may hold about something. Philosophy in the academic sense is a discipline focused on systematic and argued reflection on reality. It deals with questions such as what is the nature of reality? What kind of things can be said to 'exist' in reality and in what ways can things exist? What kind of knowledge can we have about reality? How can we think about good and evil in reality? What kind of values can we pursue in reality? In what ways can we act in reality? These different questions have led to different subdisciplines of philosophy (metaphysics, ontology, epistemology, ethics, axiology, methodology). All of these can have a value for product innovation management as will be shown in the following chapters. After all, in developing technology, we want to use knowledge, do things in a certain way, and take into account what people will consider to be morally OK or not. We will have to think about the values that customers will look for, and reflections on the nature of artefacts in general may help us recognize what aspects need to be taken into account when designing and making them.

One of the most important things philosophy can do is to help in stating the relevant questions, perhaps even more so than providing the answers. Numerous innovations have failed simply because people did not ask relevant questions but made assumptions without questioning them. We assume that customers know what they want, so we ask them what they want and we get an answer from them. Later, to our surprise, they do not respond positively to the product that we developed for them entirely according to the answers they gave us. The precious lesson we learn is that we incorrectly assumed that they would be able to state their requirements in a phase in which there was no tangible or visible product yet. Having philosophical reflections in determining a product strategy can bring such questions to the foreground and prevent naïve decisions.

In this book we will start with some fundamental issues related to the nature of the reality in which products and strategies are developed and used. In Chapter 2, the notion of 'worldviews' will be used to investigate what type of fundamental questions can be raised. There is no reason to shy away from this notion, as it is nothing esoteric by nature (Vidal, 2008), although some worldview can be rather esoteric of course. Also, it need not be associated with religion immediately, although there are worldviews of course that are inspired by religions. What the notion will be used for is to show that some assumptions about reality may be more favorable for thinking about technological developments than others. That is not to suggest that one worldview is 'correct' and another is not; it just shows that one worldview is a more 'natural' starting point than another if one seeks for guidance in the development of new products and technologies.

In Chapter 3, the social context in which products and technologies are developed and implemented is discussed from a philosophical point of view. Reflections on the nature of this context lead to all sorts of conditions for a successful and responsible innovation. We will see how this social context contains a variety of actors that all have some kind of impact on the innovation, whether we like that or not. Some of these actors may have very outspoken ideas about what the world should look like, and an extreme way to express these is in the form of a utopia – an image of an ideal world. Utopias can either hamper or stimulate innovations, depending on how we treat them.

In the chapters following these more fundamental chapters, we will follow the process of innovations more or less step by step. A caveat here is that the order of the chapters does not necessarily suggest the temporal order in which steps are taken in innovations. Different innovations can take different paths. In fact, the idea of linearity in innovations is one of the ideas that will be challenged by philosophical reflections. Yet, the phases that will feature in the consecutive chapters will be part of innovations in some way or other.

Chapter 4 will focus on the variety of conditions that emerge from a reflection of reality of which its complexity will be taken into account. This will be spelled out in a range of aspects, each of which has its own regularity and purpose. Chapter 5 will continue with methodological considerations about the many schemes and flowcharts that one can find in handbooks for innovations. Chapter 6 focuses on the role of research in innovations. This research has a more-or-less scientific character, 'more or less' referring to the extent to which outcomes may or may not contribute to the development of scientific theories or just provide practical clues for the development of products and technologies. In the next chapter, Chapter 7, the way innovations can be and are assessed and valued is discussed. This chapter will particularly look at ethical considerations and the issue of sustainability.

An important disclaimer for this book is that it does not aim at prescribing how to do things, as happens in most books about management for product innovation. Rather, it aims at discussing the sort of assumption we make and the sort of questions that we need to pose in product innovation management. In that sense, this book complements the vast amount of literature that has a more cookbook type of nature. The philosophical orientation of this book does not make it less practical, though. It shows that innovating is not just a matter of following standard guidelines and carrying out standard procedures that need no further critical reflection. Philosophy can be seen as the 'voice of the conscience'. It can be annoying to hear it express its concerns, but in the end, it mostly proves to be worthwhile to take it seriously. After all, prevention is better than cure.

References

Fives, A. (2016). The Role of Philosophy in Public Matters. In Fives, A. and Breen, K. (Eds.). *Philosophy and Political Engagement. Reflection in the Public Sphere*. London, UK: Palgrave Macmillan, 7–27.

Morris, T. (1999). *Philosophy for Dummies*. New York, NY: Wiley Publishing, Inc.

Vidal, C. (2008). What Is a Worldview. In van Belle, H. and van der Veken, J. (Eds.). *Nieuwheid denken. De wetenschappen en het creatieve aspect van de werkelijkheid* (transl: Thinking Novelty. Sciences and the Creative Aspect of Reality). Leuven, Belgium: Acco, 1–12.

2 Philosophical perspectives on innovation

2.1 Worldviews as a philosophical tool

"Every story is informed by a worldview" – this is a quote from Brian Godawa, a famous Hollywood movie screenwriter. The very same holds for management strategies in technological innovation and research. What we plan and do is somehow influenced by our basic ideas about reality and this is exactly what a worldview (or in German: *Weltanschauung*) expresses. Therefore, it makes good sense to discuss worldviews in the context of research and innovation, if only to prevent what Per Galle called 'insidious inconsistency' in design due to a lack of coherence in our assumptions about reality (Galle 2008). Philosophers have a natural connection with worldviews as they aim at developing systematic reflections on the nature of reality and worldviews are expressions of basic ideas about that nature of reality. No wonder that many philosophers have been involved in defining and describing worldviews as a tool in philosophy. Well-known philosophers such as Kant, Hegel, Kierkegaard, Dilthey, Husserl, Jaspers, and Heidegger are among those. There even exists a Worldview group of philosophers that is now called the Center Leo Apostel for Interdisciplinary Studies, named after one of its founders, a Belgian philosopher who worked at the Free University Brussels.

Apostel conceptualized a worldview in terms of six 'big questions' that correspond with six philosophical subdisciplines. Each particular worldview is a coherent set of concrete answers to these questions. According to Apostel, these basic questions are (Aerts et al., 1994):

1 What is? This ontological question leads to a model of reality as a whole.
2 Where does it all come from? This explanation question yields a model of the past.

3 Where are we going? This prediction question results in a model of the future.
4 What is good and what is evil? This axiological question is answered in a theory of values.
5 How should we act? This praxeological question leads to a theory of actions.
6 What is true and what is false? This epistemological question is answered in a theory of knowledge.

It would perhaps have been more appropriate to relate the fourth question with ethics as a subdiscipline of philosophy, as axiology is generally taken to be a wider term that also comprises aesthetical values. Therefore, in this chapter the term 'ethics' will be used here only for questions about good and evil and axiology for questions related to values in general, not just moral values.

For each of the six questions we will see how different answers lead to different perspectives on product development strategies and policies. We will start with a worldview that perhaps is one of the most elaborated ones in terms of the questions posed by Apostel and it is also the one that we will use in this book most of the times. It was developed by a Dutch philosopher, Herman Dooyeweerd. It has its roots in Christian faith, but it appeals to non-Christians as well. In the Netherlands, among other places, it has been taught to student engineers of very different backgrounds for many years and almost always evokes immediate recognition. Having seen this particular example of a worldview, we will compare it to other possible worldviews, not to give a ranking of 'better' or 'less good' worldviews, but to get an idea of their 'fit' with the needs of product development management.

2.2 The ontological question

The first question, 'what is?', according to Dooyeweerd, can be answered in a twofold way. In the first place, reality has different aspects, each of which has its own regularities (or 'laws', although this term is not unproblematic from a philosophical point of view). These aspects are 'real' in that they are not the product of our imagination but manifest themselves as we experience reality. This way of studying reality refers to what Dooyeweerd calls the 'law-side' of reality, because of the regularities than can be observed in each of the aspects. In Table 2.1 a list of 15 aspects is presented. There has been debate about this number; most philosophers do not see it as definite but rather as a number that on the one hand is large enough to provide insight into

Table 2.1 Aspects of reality in Dooyeweerd's ontology (Dooyeweerd 1955)

Name of aspect	*Relevance for products (example)*	*Regularity (example)*
Numerical (quantitative)	Number of parts	1 + 1 = 2
Spatial	Space taken by product	Sum of angles in triangle
Kinematic (motion)	Joint can move (or not)	Conservation of momentum
Physical	Material properties	Conservation of energy
Biotic (life)	Contact with living tissue	Survival of the fittest
Psychic (sensitive)	Customer perception of product	Psychological consumption law
Analytical (logical)	Reasoning about the product	Rules of logical inference
Formative (historical)	Product development process	Logic of steps in design
Symbolic (linguistic)	Name of the product	Grammar rules
Social	Impact of product on society	Politeness
Economic	Cost of product, price of product	Law of diminishing returns
Aesthetic	Appearance of product	Simplicity is beautiful
Juridical	Compliance with legislation	Private or public laws
Ethical	Safety of product	Deontological (duty) rule
Belief (trust)	Trust evoked by the product	"Trust comes on foot and goes on horseback"

the richness and complexity of reality and on the other hand can still be overseen and even turned into practical tools for management (see Lems 2008 for an example in water management and Basden 2008 for an application on information systems). The table also provides an example of the meaning of each aspect for product development and an example of a regularity or 'law'.

The first five aspects have in common that their regularities do not involve human intentions. They do not require obedience to work. From the psychic aspect onward, the aspects involve consciousness and the ability to ascribe values. In the analytic aspect, we ascribe truth values (true or false); in the formative aspect, we ascribe developmental values (progress, regress); in the symbolic aspect we ascribe symbolic values to things (for instance, by naming them); in the social aspect, we ascribe social values (friendship, for instance); in the economic aspect we ascribe economic values (cheap, expensive); in the aesthetical aspect, we ascribe values of beauty (beautiful, ugly); in the juridical aspect, we ascribe values of righteousness (just or unjust); in the ethical aspect, we ascribe moral values (good or evil); and in the belief aspect, we ascribe trust values (trustworthy, not trustworthy). Rules in these aspects do not work 'automatically' but only when we as humans choose to observe them. That makes the scientific disciplines that study them less easy because human behavior is more difficult to predict than the behavior of, e.g., elementary particles.

An important claim made by Dooyeweerd is that no aspect can be reduced to one or more of the other aspects. Some people think that gaining trust is a matter of providing good arguments. They think the trust aspect can be reduced to the analytical aspect. Or they believe it is a matter of psychology only and changing perceptions will do to gain trust. All that has proven not to work. People sometimes persist in distrusting a technology or product in spite of all the efforts based on other aspect to gain trust. Efforts to set people at ease about the safety of nuclear power plants by presenting all sorts of calculations on risks usually fail. Evidently people have a more independent reason to trust or not trust and not one derived from other aspects of reality. Trust is something 'basic' apparently. The same holds for all other aspects (in principle, but of course this depends on the way the complexity of reality is spelled out in a list of aspects and 15 is not the 'necessary' way to do that; other divisions may result in aspects that can be reduced to other aspects).

The second way of answering the question of what exists is via entities (objects and events); in the case of this book in particular, via products (artifacts) and services. In this what he calls object-side of reality, Dooyeweerd spelled out different ways in which these can function in reality. He distinguishes between subject and object functions. A house that is sold is a subject in the physical aspect because it exerts forces on the ground, and it is an object in the economic aspect. A house can never be an object in the economic aspect because it cannot buy or sell.

What alternative answers are there to the ontological question of what is? The opposite of a worldview like the one sketched above is a reductionist worldview in which reality is reduced to few or even one aspect. Most popular is the worldview that the only thing that really exists is matter. All else, such as psyche and supernatural beings, is at best a by-product of matter. The fact that we have the impression to have a free will is wrong because in the end we are a set of molecules, be it a very complex one. Thoughts and beliefs are the chemical reactions that take place in our brain. Although this worldview does not make it impossible to speak about emotions and will, they discuss them in a sort of 'as if' mode. They do not really exist, but in all our materiality we experience them and therefore feel the need to somehow take them into account. Philosopher Daniel Dennett calls this level the 'intentional stance'. But in the 'physical stance', all that appears to be an artifact of material processes in our brain. The intentional stance does not make it impossible to take into account nonmaterial aspects of reality, but it is not the most logical one to do so as all intentions are of an 'as if' nature in this view. Nonetheless, many people hold it and this group maybe even includes some of the best experts in product development. Alternative reductionist views are as follows: in the end everything is a matter of money (the economic aspect), or in the end everything is a matter of what is in the eye of the beholder (the psychic aspect), or even in the end everything is a matter of the nuts and bolts being in the proper place (the formative aspect). The latter view can be called technicism. All of these run the risk of oversimplifying the complexity of product development. Some companies have learnt hard lessons by reducing too much to technical perfection or the price tag of the product. Whatever worldview is held, the complexity of reality forces itself on us, whether we like it or not. At the same time, we should acknowledge that this complexity gives us a wealth of possibilities to optimize the product as each aspect can be seen as a 'button that can be turned' to get a good tuning of all requirements related to the various aspects.

2.3 The origin question

The second question in Apostel's conceptualization of worldviews is that regarding the origin of things. Where do things come from? In the Dooyeweerdian approach, this has been conceptualized in the foundational function of objects. This function tells us what caused objects to become. For this function there are different aspects available. Stones in the river have their origin in physical processes of erosion. Their foundational function is in the physical aspect. Trees in the wood have their foundational function in the biotic aspect as it is life processes that made them to what they are now. Technical artifacts came about by processes of human development and so they have their foundational function in the formative aspect. This way we can conceptually distinguish between raw materials and half-products. Raw materials have their foundational function in the physical or biotic aspect, while final products and half-products have it in the formative aspect.

Does it make a difference in which aspect the foundational function is? In a pure materialist worldview, it does not because natural and human causes are not fundamentally different as humans are nothing but a complex system of material elements. In a nonmaterialist worldview, there is a difference in that humans have a nonmaterial mind. That mind does not feature in physical and biological processes, but it does in technological and other cultural processes.

Are there any consequences to this difference in worldviews? A possible consequence is a different perspective on the possibility of computer- and robot-created products. If creativity is a phenomenon that has its origin in material processes in the brain, then it is possible that we develop Artificial Intelligence that in the long run will be able to display the same level of creativity as humans do. Already now there is computer-produced artwork. There is controversy about the issue of artistic value of these products. If they are not the creation of a human mind, but a machine's product, can they really be said to have cultural value? Or is the cultural value rather in the designing and programming of the computer that brought forth the artwork? Computers can support the design of products, but can they fully take over design work? Even if it becomes more and more difficult to distinguish between a human- and a machine-designed artifact, does that mean that they really are the same? For now, it seems that artificial intelligence does not have the level of originality that humans can reach. Those who have a worldview in which humans have a nonmaterial mind will claim that it can even be misleading to deny the difference between humans and machines in product development. Machines are excellent information processors, but they lack a mind as humans have it. In their view, a cooperation between humans and machines can get the best out of both and thus create an optimal condition for successful product development.

2.4 The purpose question

This question about the future can be translated to the quest for purpose in the context of product development. What is the product for?

Dooyeweerd developed the notion of the qualifying function for answering this question. This function indicates what the object is ultimately for. A bridge is there to connect two shores, but that is not its ultimate aim. This connecting function, which features in the spatial aspect, can be called its 'technical' or 'operational' function. Bridges can have different qualifying functions. The famous Chain Bridge in Budapest caused two separate villages, Buda and Pest, to become one city. This bridge has its qualifying function in the social aspect: two communities became connected. The small bridges in the

Haarlemmermeer (one of the many polders in the Netherlands) designed by the Spanish architect Santiago Calatrava hardly play any role in traffic but are a sort of sculpture in the landscape. They have their qualifying function in the aesthetic aspect rather than the social. And the bridge leading from the mainland to the Church Island in Wales only served to allow churchgoers to reach the church and thus had its qualifying function in the belief aspect (religion is an ultimate form of belief). So different bridges can have their qualifying function in different aspects of reality, and in principle that holds true for every object.

2.5 The value and moral question

Morality is part of any worldview. So are other forms of values, such as aesthetic and economic values and values of justice. Different worldviews have different perspectives on the relation between these values and reality. Some worldviews claim values are intrinsic in reality. In ethics, moral realism is an example of that view. Others claim that values are merely a matter of ascription. Something is not intrinsically good or bad, but we can ascribe moral goodness or wickedness to it. There are good arguments for both perspectives. It is clear that in many cases, different people have different ideas about the moral, aesthetic, economic or justice value of a product. This seems to point in the direction of values as being ascribed, not inherent. At the same time, we can see that some products seem to give rise to much uniformity in the appreciation of that product. A Rembrandt painting is seen as beautiful by people through the ages. Killing humans simply for pleasure is seen as immoral throughout the ages and in all cultures (even the cannibals have morality in their killing of humans).

When it comes to moral values, it is also common for all peoples and times that it is acknowledged that evil exists and that people can have evil intentions. Strikingly this is often overlooked in product development. Designers tend to focus on the benevolent users only, namely those that will use their product according to their (good) intention. The screwdriver is designed and made for people who want to drive screws with it, not to kill someone else with it. Yet, we know that the possibility exists that in all our good intentions to drive screws with it, we can suddenly be overcome by anger and rage while someone is standing in front of us and in reach of our screwdriver tip. Designers increasingly become aware that not only proper but also accidental and even immoral use of their product is a necessary consideration in product development. Since 9/11, we know that airplanes can be used for other purposes than for bringing people from A to B. Designers feel the responsibility at least to reflect on the possibility of preventing immoral use of a product (as far as this is possible) or design the product in such a way that the impact of immoral use is limited.

2.6 The methodological question

Methodology is the discipline that deals with the way through which we can do things. This description is derived from the literal meaning of *meta hodos* in Greek: 'the way along which' (we can do things). This 'way' can be highly ordered, but it can also be less predictable. One worldview is that of modernism and in that worldview control over this 'way' is very important. Modernism is based on the assumption that reality is highly controllable. Not only that, but we as humans also need to make use of that controllability. In the early times of Modernism, this was operationalized in a discipline called cybernetics. It was defined by Norbert Wiener in 1948 as the scientific study of control and communication in the animal and the machine. Those who believe in (almost) complete controllability of reality often also believe in the possibility of realizing an ideal world. This we call utopian thinking or utopianism. We find it quite regularly in the rhetoric that accompanies technological developments. This rhetoric does not so much come from engineers, as they, more perhaps than anyone else, know that the perfect product does not exist. It often comes from those who stand at some more distance from the actual product development. Politicians can justify the decisions about large investments in technology policy by referring to the need for a world without diseases (preferably even without death), a world with unlimited communication and transportation, a world in which each and every individual is served according to his or her needs. We also find this utopian thinking in advertisements. Mark the frequent appearance of the word 'unlimited' in advertisements. 'Unlimited data', 'unlimited internet', 'unlimited seafood', 'unlimited car wash', etcetera. The validity of this ideal of unlimitedness can be questioned. Do we not get tired of unlimited data or unlimited car washing? Does unlimited eating do good to our health? Yet, we seem often mesmerized by the utopia of a world in which all boundaries have fallen away.

A more fruitful way (*hodos*) probably is that we strive not for the perfect solution but for the one that deals best with imperfections. Not only are users imperfect in that they may not always have responsible intentions but also there are imperfections in what we can realize technically. Is not that what engineers learn in their education? How to make the best trade-offs in a list of often conflicting requirements? A worldview that leaves space for the imperfect is more likely to lead to successful products than one in which imperfection is seen as undesirable and in need of being overcome. Design methodologists often speak of 'wicked problems'. This is, of course, not to suggest that problems themselves can be immoral but to indicate that design problems are unruly, difficult to handle, complex. It does not help to try to take out this wicked character of design problems by making gross simplifications and idealizations. We have to deal with the wickedness of design problems, not ignore it.

2.7 The knowledge question

For knowledge, there are, in principle, two basic perspectives: either our knowledge concerns only reality as we experience it, but we do not know if that relates to any reality behind our perceptions, or knowledge does concern the reality itself, because our perceptions are somehow caused by a reality about which we can develop knowledge through our perceptions. The first view on knowledge is called an instrumental perspective and the second one is a realist perspective. Both stem from an a priori worldview, as we cannot prove or disprove that there is a reality behind our perceptions. That is a matter of beliefs about reality. In principle, there are different combinations between ontological and epistemological perspectives. One can believe in the existence of a reality behind our perceptions (we are not in a sort of Matrix like in the movie trilogy), but we cannot have the knowledge about it. This combines ontological realism with epistemological instrumentalism. One can also be a realist both ontologically and epistemologically. This seems like a fruitful starting point for product development. Of course, a product developer with an instrumentalist view can do proper work but it takes a realist view to realize that there may be a mismatch between our perception and the reality behind our perception. That attitude stimulates care about the trustworthiness of what we measure and perceive.

As we have seen, different worldviews can lead to different perspectives on the way product development and its management can be realized. One worldview may have more fruitful answers to the six questions described above than another. Therefore worldview-based considerations do matter for product development. In this chapter, various worldview elements have been presented with their consequences for view on product development.

References

Aerts, D., Apostel L., De Moor B., Hellemans S., Maex E., Van Belle H., and Van Der Veken J. (1994). *Worldviews: From Fragmentation to Integration.* Brussels, Belgium: VUB Press.

Basden, A. (2008). *Philosophical Frameworks for Understanding Information Systems.* London, UK: IGI Publishing.

Dooyeweerd, H. (1955). *A New Critique of Theoretical Thought*, Volumes I–IV. Jordan Station, Ontario: Paideia Press.

Galle, P. (2008). Candidate Worldviews for Design Theory. *Design Studies*, 29(3), 267–303.

Lems, P. (2008). The Values of Urban Water. *Proceedings of 11th ICUD – International Conference on Urban Drainage.* 31 August–5 September 2008.

Wiener, N. (1948). *Cybernetics or Control and Communication in the Animal and the Machine.* New York, NY: Technology Press.

3 The social dimension of technology and engineering

In this chapter the role of social actors in product development is discussed. This is a complex and dynamic matter that cannot be reduced to certain standard procedures for product development. First, an approach to survey the various actors with their interests and means of power is described (the actor-network approach). Then, we focus on co-design as a design approach that strives for active involvement of at least some of these actors in the product development process. This will bring in the values these actors hold that value-sensitive design is an approach to incorporate those values through conceptual, empirical, and technical analyses. Thereby scenarios can be used to sketch an integral picture of a desirable future in which new technologies function in such a way that justice is done to social and personal values. An extreme form of scenarios are utopias. Finally, the focus shifts toward the norms in product development (and the management guiding it) as a social practice in which professionals act with certain authorities and responsibilities. The matching of the norms in the practice is crucial for a healthy functioning of that practice.

3.1 The mutual relationship between society and technology

Technological and social developments are closely intertwined. There is an influence of technology on society and vice versa. This mutual relationship is something that needs careful consideration in product development. This is no news to companies that develop new technologies, but often this notion is 'translated' into procedures that are expected to enable the company to deal with this relationship in a proper way. Such procedures often try to simplify the relationship so that it becomes manageable. This can, however, lead to oversimplifications that frustrate the positive effect of the procedures, according to, among others, Geoff Crocker (2012). In this chapter, we will again call in the philosophy of technology to get to know better the nature of this relationship. Thereby we will pay particular attention to the social actors that are in some way or other involved in technological developments and the implementation of new technologies. We will focus on five issues in the philosophy of technology that shed light on the complex relationship between social and technological developments:

- The actor-network approach that deals with the different types of social actors, their interests, and means of exerting influence,
- The notion of co-design in which different actors cooperate in technological developments,
- The notion of value-sensitive design in which the different values held by actors are taken into account,
- The use of more-or-less utopian scenarios to get an overall image of the future of technology and society, and
- The normative practice approach that analyzes the responsibilities and means of different actors and the way professionals in social practices can fulfill their role properly.

3.2 The actor-network approach

The actor-network approach has its roots more in sociology than in philosophy, but it does call for philosophical reflections on the role of social actors in technological developments. Often the term 'actor-network *theory*' is used, but that suggests that it has an explanatory function, which is not the case. The actor-network is an analytical instrument to gain insight into the way different social actors hold different interests and values and try to exert power to have their interests addressed.

The relevance of the approach for product development management is that it helps to see how the company is only one actor in a whole network of social actors. Different types of actors and interests can be identified:

- The companies themselves: their primary interest is to make a profit. That is the basis for their existence. Whatever other praiseworthy goals they may have, they cannot exist when losing money in their business. In the terminology that was used in Chapter 2, they are economically qualified. Although they may also have goals such as promoting happiness and providing care for people, they are not the same as foundations, charities, and other actors who do not have to make profit because their existence is guaranteed by donations or membership fees. Their primary means of exerting power consists of the products they sell. In addition, they may participate in negotiations with governments and legislation that will have an impact on their activities, and they will communicate with customers in order to understand how to ensure that these customers will buy their products. They can also exert influence through the patents they own. These patents can withhold others from using knowledge that has been developed by the company.

- Governments: their primary interest is to safeguard public justice. Their means of power are legislation, grants, granting certain rights and jurisdiction to mention the most important. There are also government-related organizations that have an advisory or executive function.
- Knowledge-producing actors such as universities and other publically funded research labs: their primary interest is to develop knowledge based on scientific research. Their means of power are the use of the knowledge they produce. Through publications they can create conditions favorable for developing new technologies.
- Interest groups, such as environmental groups, organizations representing citizens' interests and charities: they can influence the public's behavior through dissemination of information and ide-a(l)s. They have information and money as means of power.

A more complete survey of possible social actors involved in a technological development can be derived from Table 2.1 in Chapter 2, in which a survey of aspects of reality was presented. It is possible to relate most of these aspects with social actors whose existence is directed toward that particular aspect. In Dooyeweerd's terminology (see Section 2.4), each actor is qualified by a certain aspect of reality. The reason we have governments, for instance, is because their function is to maintain public justice. A government that sees its primary role in controlling markets is not acting in accordance with its qualifying function (although this depends, of course, to some extent on the view one has on governments). Governments have a relation with markets, but only in so far as interference is needed to safeguard public justice (for instance, in the case of an organization selling human organs and thereby abusing the economic position of the poor). In Table 3.1, a survey of actors and the aspect in which their qualifying function lies is presented.

Table 3.1 Aspects, social actors and their values

Name of aspect	*Example of actor*	*Value*
Numerical (quantitative)	Bureaus for statistics	Expressing reality in numbers
Spatial	Land use (zoning) committee	Value of space
Kinematic (motion)	Transportation-related actors	Enabling people to travel
Physical	Energy-supply related actors	Providing energy
Biotic (life)	Environmental action groups	Protection of natural environment
Psychic (sensitive)	Human factors department	User-friendliness (transparency of use)
Analytical (logical)	Scientific organizations	Proper understanding of functioning of products
Formative (historical)	Engineering departments	Well-functioning of products
Symbolic (linguistic)	Language committee	Proper use of language
Social	Societal associations	Social welfare, community
Economic	Industrial company	Profit
Aesthetic	City planning advisory board	Harmonious appearance
Juridical	Government	Public justice
Ethical	Ethical committees	Morality
Belief (trust)	Church, ideological organization	Religious/ideological values

In the actor-network approach only horizontal relations are taken into account. In reality, vertical relations also exist. Some actors have authority over others. Governments have authority over citizens. Universities and companies have authority over their employees.

According to the actor-network approach, actors need not consist of humans exclusively. Nonhuman actors are also part of the 'force field'. Philosophically this raises questions. In particular, Latour has argued that technological devices can be ascribed the ability to 'act'. One of his examples is the speed bump in a residential area. This bump clearly has a power to influence road users. But is this 'acting' in the full sense? One of the most pressing questions is that or regarding responsibility (or liability in case of legal affairs). Can the bump be held responsible for possibly damaging the car in case it is not well visible? Maybe for the bump we easily deny responsibility but how about humanoid robots (or even robots in general)? They look more and more like us. Should not they be ascribed actorship and with that responsibility? Here the worldview one holds comes in again. In a materialist worldview, there is no fundamental difference between humans and robots, and in that perspective, it makes sense to think about proper ways of ascribing responsibility to robots, given the fact that we ascribe that to ourselves also. In a worldview in which the human mind cannot be reduced to matter, this is problematic and there is good reason to distinguish between human and nonhuman actors.

This is not a theoretical matter only. Computer software is sometimes seen as a partner in design work. No doubt, this software can yield impressive results in terms of finding new shapes and structures for certain functions. But it lacks the originality and creativity that a human mind has. The solutions will always remain within the (continuously extending but limited) repertoire of the program. Only in a cooperation between computers and humans can this software reach its full potential.

3.3 Co-design

Co-design is the active involvement of users in innovations. This is by no means a new concept. It probably goes back as far as the 1970s. It is now a widely used form of design work and has proven its value. User requirements before the 1970s were established mostly by the designers themselves, with market research as an input. But the best way to get to know what users really want is to get them actively involved in the design process during its whole duration. Co-design in fact focuses on two social actors in the whole actor-network, and for that reason it cannot be the exclusive approach in design, as there are more relevant actors than users only. From a philosophical point of view, it is important to notice that co-design, or co-operative design, takes a different stance toward design than traditional design methods do. Traditional approaches can also involve users but only in the very beginning, as they assume that first the voice of the customer needs to be heard by the designers, who then 'translate' their requirements into product specifications and next in product features. The customer's voice may then again be heard once the product is almost finished and a prototype is given to prospective users to investigate how they value it. Co-design takes a different stance in that the customer is also actively involved during the design process. Of course, the customer lacks the knowledge and skills to do the design work, but (s)he can bring in knowledge and skills that complement the designers'. Designers can make 'personas' to sketch different types of customers, but these can easily become abstract and ideal. Real customers may be more complex in nature and more dynamic in their requirements than suggested in a static persona. In co-design, design is not seen as an activity that is determined by designers only but in which other actors, and that goes even beyond users/customers, are partners with active participation throughout the process.

3.4 Value-sensitive design

Related to co-design, but with a different focus, is value-sensitive design. This approach emerged somewhat later, in the 1980s. Again, the interests of prospective users are at stake, but now it is primarily the values they hold, and in particular their moral values. Here again we run into the importance of worldviews for innovations. They are relevant to make explicit not only what values management holds for innovations but also how external parties hold views on values in innovations.

Often the word 'progress' is used to indicate the effect of new product development and use. But what do different actors mean by 'progress'? Progress is getting from worse to better. But what is seen as 'worse' and what as 'better'? It is striking how often answers to these questions are assumed without any systematic reflection. It is easily assumed that more intelligent is better, as is living longer and being healthier and stronger. But is that true under all circumstances? Through human enhancement we can make people stronger and more intelligent, it is believed (and probably we will indeed be able to realize much of that with new technologies). But what if an enhanced person uses the new strength and intelligence for terrorist purposes? What if we can make people live longer but the falling away of time limitation paralyzes them in their motivation to finish a study or a project? These examples show that it is not to be taken for granted what values to strive for.

Table 2.1 in Chapter 2 presented a list of aspects in reality. This list can also be used to identify different types of values people can hold in what Friedman, Kahn, and Borning (2008) call the 'conceptual analysis' in value-sensitive design. Usually values are initially taken to be economic values. These values relate to the economic aspect of reality. But this is only one of the many aspects of reality. There can also be social, aesthetical, moral, and trust values, just to mention a few other types of values. As Friedman, Kahn, and Borning point out, trust values can be particularly important for ICT systems. Empirical investigations can provide insights into the way different actors deal with different types of values. Technical investigations can then show how these values are or are not addressed by certain features of a product. A comprehensive analysis of the aspects of reality, the social actors whose existence is related to that aspect and the values they hold is presented in Table 3.1.

As Van de Poel (2018) has pointed out, it is important to be aware of the fact that values may change once a product is in use for a longer time. Perhaps unexpected and/or undesired side-effects have become evident. That may cause people to realize that they hold implicit values that they did not originally deploy to assess a product but that now get to the surface as these side-effects become visible. Also, people may become aware of new possible uses of the product and for that reason assess the product with other values than they originally used. This dynamics in values needs to be taken into account in product development management.

3.5 Utopian scenarios

One particular tool in value-sensitive design is the use of scenarios in which the way values play out in possible futures is elaborated into encompassing images of those possible futures. There is a certain trend to look for scenarios that portray an ideal, perfect, and flawless future. This scenario may not be one that is meant to be realized, as most engineers know that the perfect product simply does not exist. To the contrary, one of the most important things engineers learn in their education is to make trade-offs. The best product is not the perfect product but the one in which the best trade-offs are made. Yet, it can be tempting to try to get as close to the utopia as reasonably possible. This raises philosophical questions with practical implications for innovation management.

These questions are often worldview-related. Utopian scenarios can be driven by assumptions of total controllability of reality. This ideal of total control emerged in Modernity, with its high expectation of science and technology. Modernity has its roots in Enlightenment but became particularly strong in the USA and Europe after World War II (WWII). It was expected that the use of science and technology would enable the rebuilding of the war-stricken countries and lead to a society in which every citizen would prosper. By the end of the 1960s, this belief started waning as many people felt that science and technology had brought alienation rather than the promised freedom. This disappointment was soon followed by Post-modernism that in fact is more an enhanced form of Modernism than an abandoning of the modernist ideal. Modernism strived for control by standardization and uniformity (with the advantage of mass production) and Post-modernism shifted the level of control to individuals and pluralism. But the idea of controllability of reality was very much the same. This controllability, however, time and again appears to be more limited than expected. One particular aspect of this is the way nature responds to our actions to gain control over her. We now know that there are limits to what natural systems can endure. This makes utopian scenarios potentially an instrument for a control-based approach to reality that will appear to be a potential dystopia when reality 'bites back'.

Utopian scenarios need therefore be used as indications of directions to be taken, not as future situations that need to be realized as the scenario describes them. When using them for product development and the management supporting that, a constant awareness of the limitations of the scenario need to be in the designers' and managers' minds.

3.6 The normative practice approach

In the actor-network approach, the level of analysis is the social actor and not so much the individuals working in those organizations. These individuals do make a difference in the organizations in which they are active. Their role gets more attention in the so-called normative practice approach. In this approach, the concept of social practices, as developed by the philosopher-ethicist Alisdair Macintyre, is combined with the notion of norms. Social practices are

> any coherent and complex form of socially established cooperative human activity through which goods internal to that form of activity are realized in the course of trying to achieve those standards of excellence which are appropriate to, and partially definitive of, that form of activity, with the result that human powers to achieve excellence, and human conceptions of the ends and goods involved, are systematically extended.
>
> (Macintyre 1997)

Clearly, a business corporation is a social practice in that definition. There is a well-defined set of activities that aims at realizing a good that is often expressed in a mission statement. The standards of excellence are also often made explicit, possibly by adopting the ISO standards. In the normative practice approach, this notion of social practices is extended by pointing out different types of norms that are used in the practices. Roughly speaking, two types of norms can be distinguished: norms that define the practice ("structural norms") and norms that give direction to the acting of the practice ("directional norms"). Examples of structural norms are the authorities and responsibilities for each department in the corporation. The manufacturing department, for instance, has different rights and responsibilities than the sales department. An example of a directional norm is the following, used by a USA retail corporation: "We save people money so they can live better". Other mission statements can be related to sustainability targets.

It is important for the well-functioning of the corporation that the structural norms and the directional norms are aligned (Harandi, Nia, and De Vries 2015). It is not credible that a corporation has sustainability as one of its highest values, but no department in the corporation is responsible for it or has the authority to take measures. A particular tension that is present in organizations that hold a strong belief in the controllability of reality is that the professionals working in it are so much limited in their own responsibilities that they feel they cannot act according to the norms of their professions. An example of this is professionals working in care organizations feeling themselves crushed between economic norms imposed by the management and their own professional norm of spending time on sufficient attention for the patient as a human being, not as a machine that needs repair (Glas 2019). This tension may result in the professionals losing motivation and making mistakes (de Vries 2019).

Not only within one organization as a practice tensions between norms can occur, but also between organizations within a wider practice. This needs to be dealt with particularly when these suborganizations have to cooperate in a technological development. An example of how very wrong this can go is the Challenger explosion that was the ultimate result of disagreement between the engineers in the Morton-Thiokol company who were concerned about the safety of the astronauts in case the launch would take place at a temperature below 39°F (4°C). NASA, however, was focused on a possible loss of prestige when the launch would be postponed again. The management of Morton-Thiokol was concerned about their relation with NASA because for them this was an important source of income. They finally gave in to consent with the launch of the Challenger. Thus, a clash of norms between different social practices and within the company as a social practice resulted in a disaster.

Product development can be seen as a social practice in an industrial company (Verkerk 2019). Therefore, management needs to consider carefully if the structural norms that define the departments in which product developers' work are in line with the professional standards that are seen as important for the profession of designers. Structural norms regarding time and resources can be so restraining that designers feel they cannot function according to their professional standards. This may block their creativity and cause suboptimal designs.

References

Crocker, G. (2012). *A Managerial Philosophy of Technology. Technology and Humanity in Symbiosis*. New York, NY: Palgrave Macmillan.

Friedman, B., Kahn Jr., P.H., and Borning, A. (2008). Value Sensitive Design and Information Systems. In Himma, K.E., and Tavani, H.T. (Eds.). *The Handbook of Information and Computer Ethics*. Hoboken, NJ: Wiley & Sons, 69–101.

Glas, G. (2019). The Crisis in Professionalism and the Need for a Normative Approach. In Vries, M.J. de, and Jochemsen, H. (Eds.). *The Normative Nature of Social Practices and Ethics in Professional Environments*. Hershey, PA: IGI Global, 295–301.

Harandi, M.F., Nia, M.G. and Vries, M.J. de (2015). Water Management: Sacrificing Normative Practice Subverting the Traditions of Water Apportionment – 'Whose Justice? Which Rationality? *Science and Engineering Ethics*, 21(5), 1241–1269.

MacIntyre, Alasdair. (1997). *After Virtue*, 2nd ed. Notre Dame, IN: Notre Dame University Press.

van de Poel, I. (2018). Design for Value Change. Ethics and Information Technology. doi:10.1007/s10676-018-9461-9.

Verkerk, M.J. (2019). Management as a Supportive Practice: The Complexity of Management Unraveled. In Vries, M.J. de, and Jochemsen, H. (Eds.). *The Normative Nature of Social Practices and Ethics in Professional Environments*. Hershey, PA: IGI Global, 204–230.

Vries, M.J. de (2019). Conclusions. In Vries, M.J. de, and Jochemsen, H. (Eds.). *The Normative Nature of Social Practices and Ethics in Professional Environments*. Hershey, PA: IGI Global, 295–301.

4 Requirements and constraints in innovation design

The early phase

In the early phases of design, one of the major challenges is to develop a list or 'program' of requirements that the future product should meet. This is a challenge indeed, as reality is complex and multifaceted, and this complexity must be taken into account in the design process from the very beginning. There were times when industrial companies focused strongly on the technical qualities of the products they developed. Customer requirements were not given much attention as it was taken for granted that customers would buy the product anyway. In the 1950s and 1960s this was not yet a problem, as there were great expectations of technology as the means for setting up society after the destructive period of WWII. In that period, customers were not very critical and willing to try out new technologies. Later, however, when the economy did not flourish as before, customers did look at the products that were offered on the market in a more critical way. Also, governments started to pose conditions to new products, for instance, because the negative environmental effects of technology started to become evident. Technological developments thus increasingly became a game with many players or actors, as we saw in Chapter 3.

One of the challenges in gaining insight into the requirements of the various actors involved is that needs are often formulated in terms of solutions. One interesting example was the new idea that came up with the Rockwool company in Roermond, the Netherlands, to redevelop their stone wool ceiling plates for use as a substrate in growing tomatoes (mentioned in an unpublished master thesis in Dutch by Richard Pluymaekers, student at Eindhoven University of Technology, the Netherlands, in 2003). When asking tomato farmers for their needs, these farmers stated that it was important that the material was homogenous. The designers asked why this was so important and the answer was that the tomatoes need to grow spatially equal. The designers started to investigate how critical the homogeneity of the materials was and soon found out that it was not so critical at all. A good distribution of lighting and watering was much more important. When they presented draft material to the farmers, however, the first thing the farmers did was to break the board in two equal pieces and manually compare their weights to check if they were about equally light. The designers could explain endlessly that the homogeneity of the material was not important for the realization of their actual, underlying need, that is the spatially equal growth of the tomatoes, but the farmers insisted on homogeneous material and at last the designers gave up and put homogeneity in the list of requirements. Had they had the freedom to drop this requirement, other requirements could have been realized better. Had the farmers had more insight into the difference between needs and solutions, they would have left it to the designers to find the optimal physical realization for addressing the real needs.

This example shows that it is important to distinguish between needs and solutions. In the next section we will try to get a more detailed understanding of this difference. We will call in philosophy of technology, as we constantly do in this book, to find this understanding. We will do so in two steps. First we will use the 'dual nature of technological artifacts' approach to get a better understanding of the difference between needs and solutions. Then we will move on to get a more detailed view on each of these two.

4.1 The dual nature of technological artifacts

In the philosophy of technology, one of the ways to conceptualize an artifact is to regard it as an object with two 'natures', that is two ways of existence (Kroes and Meijers, 2006). In the first place, the artifact exists in a physical way. It has numerous characteristics that are inherent in the artifact and do not depend on what we think of it. The analysis of the aspects of reality as it was introduced in Chapter 1 helps to see the variety in these aspects. These physical properties are the result of choices made by the designers to reach a solution that addresses the needs of the users. The user needs form the other 'nature' of the artifact. Users may find the artifact to be functioning well, good prices for quality, easy to use, nicely looking, etcetera. These are all features of the artifact that are not inherent in the artifact but are in the eye of the beholder. For that reason, they can differ between users. One user may find the appearance beautiful and the other user may find it ugly. In the meantime, the physical characteristics of the artifact (shape, color, texture) have not changed, as they are inherent in the artifact. The second type of nature is called the 'functional nature' in the dual nature approach. It is important to note that the term 'function' must be interpreted in a wide sense. It entails all qualities of the artifact that are not inherent but ascribed. Function is one of these qualities. The object I have bought as a 'screwdriver' can also be ascribed the function of 'tin can lid opener'. That was not the function the designer had in mind, but function is not an inherent quality so users may ascribe alternative functions. Likewise beauty or ugliness, fairness of price, ease of use are all ascribed, not inherent.

In the management of design and research for design, it is important that the difference between the functional and the physical nature is taken into account. A list of requirements should contain elements of the functional nature, not elements of the physical nature (except in cases as with Rockwool when users insist on having a solution and not the underlying need in the list of requirements). Having elements of the physical nature in the list of requirements limits the options designers have to find a solution (that is, a physical realization) that addresses the real needs as good as possible.

4.2 The variety of elements in the physical and the functional nature

In Chapter 2, the complexity of reality has been spelled out in terms of 15 aspects of reality and some initial remarks have been made about their meaning for products. We will use this survey to get a more detailed view on the elements in the physical and functional nature of the artifacts-in-design. We will start with the functional nature, as it seems most logical to begin with needs and only then move to solutions. When we look at the list of 15 aspects of reality, we can recognize a certain order. The 'lower' six aspects are all 'non-intentional', that is, they do not involve human intentions. The number of legs of this table is four, irrespective of what I think of it. The size of the table is what it is. When the table wiggles, that is not because I ascribe that property to the table. Neither is the weight of the table nor the force it exerts on the ground dependent on my judgment. The table is not living, but if there was some life left in the wood of which it was made, that, too, would not depend on whether or not I believe it lives. That is different for the 'upper' nine aspects. They all involve human intention. Without intentionality, no perception, no language, no development in history, no social life, no economy, no aesthetical appreciation, no justice, no ethics, no faith or trust. This is, of course, not by accident. The Dutch philosopher Dooyeweerd, who came up with this list, made it such that each aspect assumes the previous ones

(Verkerk et al., 2016). Without 2 and 3 (the numerical aspect), no 2D and 3D (the spatial aspect). Without space to move in, no motion. For the 'higher' aspects this order is not always clear. Does language need a society or vice versa? But even though there may be some speculation on the order of the 'higher' aspects, there is merit in this order in that it nicely separates the aspects that do not and those that involve human intention.

The first aspect in Dooyeweerd's conceptualization of the complexity of reality to involve intentionality is the psychic aspect. For this aspect, the constraints it brings about are primarily in the domain of human–artifact interfacing. Issues such as user-friendliness and accessibility are key here. The work of Donald Norman is well known for its emphasis on the mental representations users make of artifacts and that can sometimes conflict with the physical realization of the product and thus cause problems in understanding the functioning of the device (Norman 2013). The artifact should be recognizable as to what it is for and how it should be operated.

The next intentionality-involving aspect is the analytical aspect. Reasoning is the keyword here. Users make logical inferences about the artifact. A sharp edge means enabling cutting. A round surface means the object can roll. Two ways of reasoning can be distinguished in particular: structure–function (or means–ends) reasoning and cause–effect reasoning. The first type of reasoning connects physical and functional natures of the artifact (this is the way the examples above have been phrased). This is a very open-ended way of reasoning, as function and structure cannot be deduced from one another. Cause–effect reasoning is, in principle, more closed. The effect of switching on the device is, in principle, deducible from the structure of the device. The examples above can be rephrased in those terms: the edge is sharp, so moving it along a surface will cause a cut in that surface. The object is round, so if given a push, it will start rolling. The constraint produced by this aspect is that the design should allow for these types of reasoning.

Then there is the formative aspect, sometimes called the 'developmental' or 'historical' aspect. This aspect calls attention for the 'logic' in the development of the artifact as one in a whole line of development. The French philosopher of technology Gilbert Simondon even conceptualized an artifact as something that is continuously in development rather than a static object (Simondon 2017; see also de Vries, 2008). He used the term 'concretization' to indicate that in the line of historical development, the artifact more and more realizes the abstract constraints in its evolving concrete form. This 'logic of development' puts constraints to both designers and users. Designers should take into account what users have become accustomed to in their experiences with predecessors of the artifact-in-design. If the artifact is a 'disrupting' technology, this poses particular challenges to both designers and users. This aspect is particularly important for the management of the product development process also. This process should reflect the 'logic' in the development, and it is in management that this logic in the overall process is guarded.

There are various ways in which the symbolic aspect puts constraints on the design. Most obvious perhaps is in the name-giving of the artifact. Naming an artifact a 'screwdriver' immediately communicates what the artifact is for (at least, when it is solid and not a colored liquid). But the symbolic aspect entails more than language. Various types of symbols can be used in the design. Architecture is perhaps the most prominent domain here. A Gothic cathedral uses many symbols to communicate its religious references (the cross-shaped floor plan, the upwardly pointed arcs, the richly colored stained-glass windows, etcetera). But also the artifacts people use in daily life can have symbolic references. Green refers to 'OK', while red refers to 'take care'. Blue refers to cold and red refers to warm.

The social aspect involves the way the product will fit (or not fit) with the way people live and work in social relations. There are examples of products that failed because of the social controversy they caused. Google Glass is a striking example of that. French sociologist of technology Bruno Latour in his book *Aramis or the love of technology* presented a wonderful description of how an innovative idea for a new metro system in Paris was crushed between the interests of different parties (Latour, 1996). It was not possible to find a design that fitted with all these interests at the same time.

The constraints put by the economic aspect are obvious. Market considerations are extensively discussed in other literature so it is not necessary to be elaborate here.

In the aesthetic aspect, the concept of 'beauty' is key. 'Harmony is the key value', as we saw in Chapter 3. This harmony concerns not only the artifact in itself but also the harmony with its environment. The artifact should not be in conflict with the context in which it will function.

The juridical aspect refers to legal constraints for the design. There are several issues here. First, there are intellectual property issues. The design either needs to avoid using patented ideas or the company will pay licenses. Second, there are environmental issues. Materials may be banned because of their problematic environmental effects. Third, there are issues of safety, privacy or human dignity that are turned into legislation. Here is a relation to the next aspect, the ethical aspect. These aspects are distinct. Some moral constraint may be difficult to transform into legislation. Legislation hopefully reflects what society considers to be good morality, but there can be tensions.

Finally, there is the trust aspect. This may involve ideological and/or religious sensibilities that have to be taken into account when designing the artifact. But it also entails the wider issue of trust in the product. This aspect can hardly be overestimated. The introduction of consecutive generations of wireless telephone technologies have been hampered because industries underestimated it. Gaining trust is not unproblematic. We know that distributing a brochure that explains why the technology is not dangerous does not have much effect. Trust must be gained in what we call 'upstream engagement'. Here we clearly have a relation with the social aspect.

All these aspect-related constraints are interrelated. The symbolic aspect is related to the psychic aspect in that symbols must be recognizable. It is also related to the logical aspect in that symbols should allow for proper reasoning and not cause incorrect inferences about the artifact. It also relates to the economic aspect. Gold parts perfectly symbolize wealth, but they come with a price. In principle, every aspect is somehow related with all other aspects, which makes the list of requirements a very complicated matter. That is why methods have been developed to get grips on this complexity (such as QFD; see Chapter 5).

The ideal product would meet all of the constraints coming from these aspects. That product obviously does not exist and never will. A survey of all the constraints will therefore only be useful if combined with other analyses that enable to set priorities in the constraints. One of these analyses concerns the different functions of the artifact-in-design.

4.3 Different functions

Let us take a pacemaker as an example.

Usually there is one aspect that contains the ultimate goal for which the artifact is designed. This ultimate function is what primarily qualifies the artifact as an entity in reality. Therefore, we can call it the qualifying function. The constraints coming forth from that function and the related aspect will have priority in the making of trade-offs in the overall list of requirements. For the pacemaker, this ultimate function is in the ethical aspect (in Dooyeweerd's approach to reality). That is because a pacemaker ultimately should contribute to the 'good life' of the patient. Care is a primary value in that aspect and that is what the pacemaker provides. It is of primary importance that the pacemaker fulfills that function. If it does not, the device does not make sense.

There is also what we can call the 'technical' function of an artifact. That is what it does very basically. In the case of the pacemaker, it is producing an electric pulse (which is in the physical aspect of reality). This function also needs to be fulfilled because without that no higher function can be fulfilled either. If the pacemaker does not produce the correct pulse, it will never contribute to the wellbeing of the patient. Note that this function is in a much lower aspect than the qualifying function.

There is a difference between the qualifying and the technical function in that the artifact is an object in the qualifying function, while it is a subject in the technical function. Whether the qualifying function is fulfilled is not up to the artifact but up to the user. (S)he either does or does not experience a contribution to wellbeing by the device. In the technical function, however, it is the artifact that determines whether or not the correct pulse is generated, and this is not something 'in the eye of the beholder', as is the case with the qualifying function. This difference needs to be taken into account when determining the constraints to be set for the device. For all aspects, the designers should determine in what ways the artifact is active in itself (for instance, in taking a certain space) or if the object is of some other active agent (for instance, in the legal aspect that prescribes when it is accepted as safe).

The device can only be a subject in the lower aspects. In all aspects it can serve as an object. There is always some aspect other than those related to the technical and the qualifying function in which the constraints are more important than for other aspects. In the case of the pacemaker, there is a very important constraint in biological aspect, in which it functions as an object. If the device is rejected by the human body, the pacemaker cannot fulfill the qualifying function. Functions in other aspects are more 'negotiable'. For instance, the economic aspect is relevant but the price of the pacemaker can to some extent be adapted to other constraints.

4.4 A systemic approach

Another way to get grips on the complexity of the list of requirements is to approach reality from a system's perspective. In a worldview in which randomness reigns, there is not much reason to take this approach. All events are unconnected and relations are purely incidental.

In most worldviews, however, reality is seen as a whole in which all elements contribute to that whole in some way or other.

One example of such a worldview is the one that was elaborated by Herman Dooyeweerd (see Chapter 2). He described two different ways in which elements can be connected in a system. For each of these there are different consequences for the requirements to the product-in-design. In the first type of connection, one is part of the other, but its function has not become dependent on the larger system. An example of that is a chip in a computer. It fulfills a certain function (for instance, amplifying a signal) but can just as well fulfill that same function in a different type of device. Similarly, a canteen in a military basis can function just as well in a factory. In this type of connection, the requirements set for that element need to be stated in terms of the requirements for the whole system. The only thing to be taken into account is the connectivity of the element to the larger system. If the chip does not fit in one of the slots of the computer motherboard, it cannot function, of course. In the second type of connection, one is part of the other and its function only makes sense in that systemic context. An example of that is a computer mouse, which only makes sense when connected to a computer. Hooking it up with a coffee machine or an mp4-player, if at all possible, does not make any sense. In that case, the requirements set to this element necessarily depend on the requirements set to the system as a whole. The third way, not mentioned by Dooyeweerd but equally important, is juxtaposition. In that case two (or more) elements exist next to each other and there is no hierarchy. An example of that is a system consisting of the Internet with different computers connected to it. The requirements for the different computers can be set independent from each other. Of course, the relation of the computers to the Internet system is different, namely of the first type (the computers do not lose their meaning when detached from the Internet).

In setting the total list of requirements, it is important to take into account what type of connection between the elements in a system is at stake for each element. This does not only hold for material systems. It also holds for socio-technological systems in which (groups of) people are connected. From an organizational point of view, it is important to analyze how different groups of people are connected in the overall organization. This determines the extent to which they can act autonomously or must obey rules that are directly related to the dependency of the meaning of their task in the whole organization. Coming back to the canteen example, the canteen personnel is relatively free to determine how they make coffee and what price they set.

But a soldier cannot function outside the military context and thus is limited in determining his/her ways of acting. This example shows that for management, an analysis of the different relations between parts of the company is important for determining the requirements to be set for that part of the organization when setting it up.

References

Kroes, P.A., and Meijers, A.W.M. (2006). The Dual Nature of Technical Artefacts. *Studies in History and Philosophy of Science*, 37(1), pp. 1–4.

Latour, B. (1996). *Aramis or the Love of Technology.* Harvard, MA: Harvard University Press.

Norman, D. A. (2013). *Design of Everyday Things: Revised and Expanded.* New York, NY: Basic Books.

Pluymaekers, R. (2003). *Sleutelprocessen bij Rockfon: QFD toegepast op het productieproces van plafondpanelen.* Master thesis Eindhoven University of Technology, the Netherlands.

Simondon, G. 2017. *On the Mode of Existence of Technical Objects.* Translation by Cecile Malaspina and John Rogrove. Minneapolis, MN: Univocal Publishing. The French edition 2012. *Du mode d'existence des objets techniques.* Paris: Aubier.

Verkerk, M.J., Hoogland, J., Stoep, J. van der and Vries, M.J. de (2016), *Philosophy of Technology. An Introduction for Technology and Business Students.* New York, NY: Routledge.

Vries, M.J. de (2008). Gilbert Simondon and the Dual Nature of Technical Artifacts. *Techné*, 12(1), 23–35.

5 Methods in innovation design research

Developing the product

In this chapter, philosophical considerations will be used to investigate how each product development in some way is unique and yet a typology can be developed to get grips on the variation of design situations. It will be argued that different situations call for different methods and that the (often implicit) assumptions underlying each method largely determine where a method can be useful and where it will hamper by hiding relevant aspects of the situation rather than doing justice to them. Many textbooks on product development treat methods as standard cookbook recipes that can be applied in almost any situation. In this chapter, a more nuanced picture of methods will be presented.

5.1 Introduction

In design methodology, initially there was optimism in identifying general design methods that would work for each and every design process (Cross 1993). Still, textbooks for design education have flowcharts of design processes with a prescriptive nature and of a generic character. Design processes, however, appeared to be more product- and context-specific than had been anticipated, perhaps because of the fact that design methodology emerged from the practitioners' experiences more than from philosophical reflections on the nature of design (De Vries 1993). Of course, the practical experiences are of great value for determining what works and what does not work in design and the management of design. But philosophical reflections have shown to be a very useful complement for those practical experiences. Such reflections point out that all these flowcharts for design processes always present a reduced and simplified version of reality. Reality is always more complex than any flowchart suggests. This complexity can be seen as problematic, but it can also be seen as a richness that one can exploit. Each process offers its own possibilities to lead to a successful outcome. But problems can easily emerge if we ignore the differences between the flowcharts as a model of reality and reality itself. Models are useful but limited. As long as the limitations are taken into account, they serve a useful role. In this chapter, we will show how different product development processes can differ due to different contexts that require adaptions of the model-like handbook schemes and flowcharts.

So due to this philosophical input, more attention was paid to the differences between different design processes. This also applies to the design research projects that are done to acquire the knowledge that is needed to come to a successful product. Most literature only describes the developmental part of product development processes, but there is also a research part. For that part, too, there is no standard procedure and each research project has its own specific goals and methods. This has led to two new ideas. The first idea is that a typology, taxonomy or classification of product development research processes is needed as there are different processes for different designs. The second idea is that these differences are related to the assumption about what knowledge is available that is inherent for the design methods. In one context, the knowledge assumption underlying a design method may be fulfilled and the method will work, but in other contexts, at least some of the assumptions may not be fulfilled, which means that the method will have to be adapted to these different contexts. These two ideas will be elaborated in this chapter.

5.2 A classification of product development processes

Zwart and De Vries, based on an analysis of master's projects at a university of technology, have developed a classification for innovative design projects. This classification consists of six project types that form the building blocks of an overall product development project. Different product development projects will contain different building blocks projects. These building block projects will now be described one by one.

5.2.1 Descriptive knowledge projects

This type of project is what most people will think of intuitively when the term 'research' is mentioned. It resembles most closely the research that is done in natural sciences. In such research, description of an existing state-of-affairs is the aim. The aim is not (yet) to assess that state-of-affairs. The physicist studies electrons and if (s) he finds one of which the behavior deviates from what an electron normally does, the researcher will not conclude that the electron is malfunctioning or broken, but that apparently it is not an electron. The aim of studying electrons is to describe their behavior and properties and not to assess that behavior by comparing with a standard for electrons. This is contrary to an engineer who studies a drilling machine and who does have an opinion on that machine. While the physicist does not talk about good or bad electron, the engineer does talk about good and bad drilling machines. For descriptive knowledge projects, no standards for the object of study need to be set. Standards for the way knowledge is gained do need to be stated. Therefore, in natural sciences, the so-called 'empirical cycle' has been developed in which new knowledge is acquired in a cycle of hypothesis development and testing these hypotheses. The hypothesis that stands against all tests will eventually become a theory and a theory that survives for a very long time eventually becomes a paradigm. The criterion for rejecting or accepting a hypothesis is 'truth', a rather complicated notion that is usually taken as 'in accordance with the observations'. Natural science hypotheses, even when they have gained the status of a theory, are still tested over and over, because there is always the possibility that a new observation will cast doubt over the truth of the theory. After all, even classical mechanics was falsified although it had survived for more than two centuries. In product development, however, a more pragmatic approach is used. The purpose of product development is not to contribute to science by delivering knowledge that will survive even in the most bizarre testing conditions, but it only needs to be valid in normal circumstances, namely those under which the product will function. The extreme generalization that is aimed for in natural sciences is not an aim here. So even though this first type of project closely resembles the empirical cycle in natural sciences, there is an important difference. Finding 'truth' is not the ultimate aim; rather, the aim is finding knowledge that functions under those circumstances under which the product will be used. Particularly in large corporate laboratories, researchers can be tempted to behave like natural scientists. The history of the Philips Research Laboratories, to mention just one example, is full of such cases. Good project management will ensure that researchers in the corporate laboratory will not fall into the trap of striving for too robust a theory.

5.2.2 Design projects

Design projects can almost be seen as the opposite of descriptive knowledge projects. Here, normativity is all over the project. This type of research aims for knowledge about requirements and functions. Both are normative in nature. Obviously, this holds for requirements because they are stated in terms of what the product *should* be like. The notion of function, however, is also normative in spite of the fact that it is mostly phrased in descriptive terms. When we say that what we develop is a car, we mean to say that it is something that *ought to* bring us from A to B. Function is what the product ought to do or enable. That still holds when the car is in the garage and cannot even bring from A to B. Still, its function is to be a car and that is why we call it a broken or a malfunctioning car. Also, the design drawings are not descriptive because here is nothing yet to be described. They are normative in that they tell what the product *should* look like once realized. The same holds for a system diagram. That, too, tells us not what the product components are and how they interact, because there are none yet, but what it *ought to* be like once the product is there. Normative knowledge is developed in a different way than descriptive knowledge and this must be taken into account in the design research management. Normative knowledge can only partially be the outcome of experiments. If we want to decide what material to choose for a part that will be stretched during its use, experiments can tell us what force will make it break, but not how much we should stay below that force in order to ensure safety. Such decisions cannot be deduced from an experiment but need consensus through debate. Such discussions need to be organized and there is a danger that the need for these discussions is overlooked because people assume that experiments are sufficient to make the decision.

5.2.3 Engineering means-ends knowledge projects

Means–ends reasoning knowledge is closely related to knowledge that is developed in design projects, in a similar way like cause–effect knowledge is related to descriptive knowledge projects. This knowledge usually takes the following shape: if you want A, and you believe you are in situation B, then you ought to do X. In that phrase, A refers to the function of the product, that is the change in reality that the product should result in, and X is the working or functioning of the product. Note that again there is an ‘ought to’ in this knowledge. Therefore, it should not be confused with cause–effect reasoning knowledge that takes a different form: if one performs X in situation B, then A will happen. In order to get to know if X is what we ought to do, we have to assess the outcome A: is it what the product ought to result in or not? This is again a matter of consensus, not of deduction. Both cause–effect and means–ends reasoning are needed in product development and need to be arranged. Means–ends reasoning is needed to get to a prototype and cause–effect reasoning can predict what will happen when we test the prototype, assuming that all the properties needed for this prediction are in the description of the prototype. Note that in the description of this knowledge, context plays a role. What is good in one situation may not be good in a different situation. Note also that contextuality is in the description of cause–effect reasoning as described here. In the first type of research project, we already noted that in product development we strive not for generic knowledge but for knowledge that holds for 'normal' circumstances, namely for those circumstances in which the product is normally used.

5.2.4 Modeling projects

Modeling projects are those projects in which the resulting knowledge is not expressed in a theory or reasoning, like in the previous three types, but in a model. Modeling can serve different functions. In the management of this type of project, it is important to make sure that the model contains what is needed to fulfil its function qua model, not more and no less. Models can be both descriptive (like in the case of the model of an existing building that is displayed in a museum for architecture) or normative (like in the case of a mock-up of a device to see if potential clients like the shape). Already the decision to use a descriptive or a normative model needs to be taken consciously. Also, the requirements for the model need to be considered carefully. A descriptive model is only satisfactory if it produces correct and relevant predictions. A normative model is satisfactory if it allows for or contains the assessment of a certain outcome. Efficiency is always to be strived for and models need to contain only what is needed to make them work. A model of a block of houses that is used to predict how shadows will fall in the course of a day need not be colored, as the color of the houses is not relevant for the casting of shadows. Models are used in different phases of the development process. In early phases, they can be used for exploring possible solutions, and in later phases, they can be used to test a chosen solution and can also serve communicative functions, such as demonstrations for clients or for educational purposes. What is left out from reality and what is taken up in the model depends on the function the model needs to serve. The list of aspects of reality that we encountered in earlier chapters can help here to reflect on these decisions. Full reality has all the aspects and every model leaves out elements of the various aspects. The function of the model is in one of the aspects and therefore elements from that aspect are less likely to be left out than elements from other aspects.

5.2.5 Technical optimization projects

These are projects that aim at improving existing projects. Clearly this type of project deals with normativity in the development of knowledge. Improving is a normative term that refers to the difference between one state (the initial product) and a later state (the 'improved' product). Methodologically this means that comparison plays a key role in these projects. Not the absolute values of parameters that describe functioning counts but their relative value compared to the original parameter values. An important first step in this process is to decide which parameters will be taken to be indicative of improvement. What decides whether or not the improved model of a car is improved? Is it its increased maximum speed? Is it its fuel efficiency, its safety, its comfort, its price or all of these at the same time? In redesigning the product, the same methods can be largely used as in the design projects. Optimization projects are an example of evolutionary development in product design. This is contrary to revolutionary design in which a new beginning is made in a line of product development. The transition from propeller engines to combustion engines in aircraft design is an example of a revolutionary design. The optimization of combustion engine–driven aircraft is the follow-up phase of evolutionary development. In terms of product innovation design research, revolutionary developments entail entirely new knowledge to be gained, while evolutionary development research is the further extension of existing knowledge. This may have great consequences for investments in research equipment.

5.2.6 Mathematical projects

The differences between this type of project and the preceding types is that it is not empirical. This particularly distinguishes it from the modeling projects in which physical models are developed and used. All models are simplified versions of reality, and in mathematical models particularly, the materiality of reality is left out and replaced by abstract parameters that describe this materiality. In current product development, the mathematical modeling takes place mostly by using computer software. A Finite Element Model in which forces in a material are predicted is an example of that. Much of what holds for modeling projects also holds for mathematical projects. The high level of abstraction in mathematical modeling offers unique opportunities because it is not limited by boundaries set by materiality. At the same time, it should be kept in mind that in reality these boundaries will again be present.

These six types of innovation research projects are distinct in aim and methodology, as we have seen. In the management of a product innovation process, decisions will have to be made about what type of project needs to be set up and in what phase of the development process. In these decisions, the nature of the product-in-design has to be taken into account. The more innovative the product, the more new knowledge needs to be gained to support the development. But even in cases of well-known products and technologies, each of the project types may be of interest to consider. The selection of project types and their sequences determines how the development process will be shaped and managed.

5.3 Assumptions in design methods

Different product design processes require not only different (combinations of) project types but also methods for design that need to be adapted to the specific conditions of each product development process. Although textbooks present 'standard' methods for design, in their application they need to be assessed in usability, based on the specific situation in which they will be deployed. All methods assume that certain knowledge is available and if this is not the case, the way the method is applied needs to be adapted to that situation, or the missing knowledge needs to be acquired through design research of some type (not only the six innovation project types, but also market research or psychological or economic research). An example that can illustrate this is the Quality Function Deployment (QFD) method that quickly rose in popularity in the 1980s and was then often overestimated in its possibilities (not in the least because the hidden assumptions were overlooked); later, the use of the method became more limited to those situations in which it worked best (not in the least because in those situations the assumptions of the method fitted the circumstances). Although the method nowadays is not as popular as it used to be, it can still serve as a rich example of how assumption in methods determine how they can or cannot be applied.

QFD aims at translating customer requirements into technical specifications. The heart of this method is the 'house of quality', a matrix in which the rows are filled with the customer requirements and the columns with the technical specifications and in the cells a score is given for the relation between these two (often ranging from 9 for a very strong relation to 0 for no relation) (Clausing and Hauser 1988). When the scores are added per row (if useful first multiplied with a factor that indicates its importance), the total score per column indicates the priorities that should be given to realizing the optimal value of that technical parameter: higher scores indicate parameters that strongly relate to important customer values while lower scores indicate parameters that relate weakly to less-important customer requirements. This method is full of assumptions that may or may not be fulfilled (De Vries 2009). In the first place, it assumes that we know who to consider to be a customer. Is it the end user? Is it the agent that will buy the product? Is it the agent that will maintain it? In the example of a touring coach, is it the owner of the coach, the travel agent that hires it, the tourist that is transported with it, or the garage where it is maintained and repaired? All of these agents have their own requirements and rarely do they all match. The method does not give any clues as to who is to be regarded the customer. Second, the method assumes that customers can express their requirements. But can they? Probably they can in cases of a product they are already familiar with, but they will have difficulties doing so in cases of very novel products. Still, the method requires input and it is tempting to fill in what customers tell in spite of the fact that these requirements may be ill-informed. Third, it assumes that we know what requirements are more and less important. Fourth, it assumes that we know the relation between customer requirement and technical specification. That, too, can be particularly problematic in cases of very novel products. Obviously for each of these assumptions, research can be done to fill knowledge gaps. But this research will only be initiated once the need for that research has become evident. It is important in product development management to make sure that discussions about assumptions in methods are organized and that research is done when needed.

QFD is used here only as an example and not because of any particular importance of the method itself. Similar considerations as presented here have to be made for any design method that is used. In particular, methods for forecasting and back-casting depend on many knowledge assumptions that need to be carefully considered before relying on the outcomes of such methods.

5.4 The use of analogies

This chapter on methods in innovation design research could be endless, as there are many methods available. The quality management methods itself is a large set and QFD is only one of them. One more method will be discussed here because philosophical reflections provide useful insights on it, and that is the use of analogies. Analogies have played an important role in science. Maxwell's formulas for electrodynamics were derived by drawing an analogy between a mechanical force system (the aether) and the electromagnetic forces system. Many other examples can be mentioned of such analogies. In design, analogies can also fulfill a useful function. But for optimal use of analogies, insights into the different types of analogies that can be distinguished is needed, and philosophy can provide those.

In the philosophy of technology, various types of analogies have been identified that can be used to enrich product development (Sarlemijn and Kroes 1988). Analogical reasoning can be a powerful tool in product development. An example of that type of reasoning can be seen in biomimicry in which analogies between nature and the product-to-be-designed are used to find innovative solutions to design challenges (Kennedy 2017). Most designers probably tend to focus on form analogies: the form of a natural object is copied because the function it fulfills is also present in the product-to-be-designed. The form of a termite hill, for instance, is used for designing air ventilation for buildings. Form analogies are only one type of analogy that can be used. Two other types are function analogies and system analogies.

Function analogies refer to copying not forms/shapes but functions. Nature can do things that surprise and intrigue us, and after some reflection, appear to be inspirational for new products that can fulfill functions that we have never before thought of. The gecko walking along the ceiling can inspire to reflect on situations in which it can be useful to walk along ceilings or walls. In this case, the shape was also copied from the biological example, but this need not be the case. There may be other forms that realize the same function that was copied from nature.

Third, there are systems analogies. Nature can offer examples where the structure of a complex object is very efficient for realizing the main function of the system. The way the various components of the eye are organized and positioned was instrumental in the design of cameras. Here we see that the analogy of the system did not lead to copying the shape also, because human-made solutions for parts of the system already existed. In the early efforts to design a transistor based on the tube amplifiers, the designers copied the layout of the triode tube into solid-state material assuming that the function analogy and the form analogy were one-to-one related. They quickly found out, however, that to realize the function analogy, a different form was needed. This historical example illustrates that function and form analogy are in principle independent, although in practice they often match.

The use of analogies illustrates that in design and design management, it makes sense to use reality in its full richness and complexity. That is in fact the basic message of this whole chapter and it relates to what was stated in Chapter 1 about the relevance of views on reality for product development. A reductionist view on reality will easily miss out opportunities and make naïve mistakes. A view in which variety and complexity are included will cause a broad view on product development. Doing justice to the nature of reality in the end always pays off.

References

Clausing, D., and Hauser, J.R. (1988). The House of Quality. *Harvard Business Review*, 1–13.

Cross N. (1993). A History of Design Methodology. In de Vries M.J., Cross N., and Grant D.P. (Eds.). *Design Methodology and Relationships with Science*. NATO ASI Series (Series D: Behavioural and Social Sciences), vol 71. Dordrecht, The Netherlands: Springer, 15–27.

Kennedy, E.B. (2017). Biomimicry: Design by Analogy to Biology. *Research-Technology Management*, 60(6), 51–56.

Sarlemijn, A., and Kroes, P.A. (1988). Technological Analogies and Their Logical Nature. In Durbin P.T. (Eds.). *Technology and Contemporary Life*. Philosophy and Technology, vol 4. Dordrecht, The Netherlands: Springer, 237–255.

Vries M.J. de. (1993). Design Methodology and Relationships with Science: Introduction. In de Vries M.J., Cross N., and Grant D.P. (Eds.). *Design Methodology and Relationships with Science*. NATO ASI Series (Series D: Behavioural and Social Sciences), vol 71. Dordrecht, The Netherlands: Springer, 1–14.

Vries, M.J. de (2009). Translating Customer Requirements into Technical Specifications. In Meijers, A.W.M. (Ed.). *Handbook Philosophy of Technology and Engineering Sciences*. Amsterdam, The Netherlands: North Holland (Elsevier), 489–512.

6 Innovation, product design, and research

6.1 The relation between scientific research and technological innovations in industry

In this chapter, the relation between scientific research and technological innovation will be discussed. In the philosophy of technology, the relation between science and technology has been reflected on for about as long as this discipline exists. The reason why the philosophy of technology emerged much later than the philosophy of science may well be that technology was long seen as merely the application of science, not as a body of knowledge of its own right. Technology was the practical follow-up of science, and as such less worth reflecting on in philosophy. The emergence of the philosophy of technology, however, soon began to question this assumed relation between science and technology. It was not difficult to refute this idea with historical references. At least two examples could be used for this. The relation between thermodynamics and the development of steam engines was certainly not in accordance with the 'technology as applied science' paradigm. Steam engines were developed before a proper understanding of what happened with the steam in the engine was established. Likewise, telescope and microscope lenses were not the application of geometrical optics; rather, geometrical optics was developed based on the need to get a better understanding of what caused optical aberrations in lenses.

The idea that technology is merely applied science became very popular in industrial corporations in the 1950s when large companies refocused their corporate laboratories to have developing 'fundamental' knowledge as their primary task. That knowledge could then by transferred to departments in which products were developed, based on the fundamental knowledge that had been gained by the central corporate laboratory. Companies such as General Electric and DuPont in the USA, and Philips Electronics in the Netherlands adopted that approach and for two decades this approach determined the company's strategy. Looking back at the two decades in which this vision was applied with Philips Electronics, it was a difficult period for the company as a whole. There were constant frictions between the corporate laboratory and the product divisions (De Vries 2005). The corporate laboratory sent a constant flow of initial 'fundamental' knowledge applications into the product divisions, but to them it was clear that market considerations raised important questions about the desirability of these applications. The perhaps most extreme example for Philips Electronics was the Stirling or hot air engine that thermodynamically was extremely interesting for the researchers, but no product divisions saw any useful application for this device. The tension between the corporate laboratory and the product division are a symptom of something being wrong with the linear view on product innovation, as if fundamental research has to precede applied research and then application of the knowledge in product development. We can understand the underlying cause of this by examining philosophically the nature of science and the nature of technology.

In Chapter 2, a reductionist view on reality and a view of reality as having many different and irreducible aspects have been presented as opposites, and it has been argued that for product innovation the latter is more fruitful than the first. For science, however, it is necessary to reduce and focus on one or some aspects. Only by abstaining from the full complexity of reality and taking out one or some aspects for in-depth investigation, science can function properly. This 'taking out' is what we call 'abstraction' (literally: peeling off). Thus, the physicist isolates the physical abstract for scientific study, the psychologist does that for the psychic aspect, the economist for the economic aspect, and etcetera. The more application-oriented the scientific discipline is, the more aspects are taken into account, but it will always be a selection and never reality in its full complexity. Environmental science, for instance, takes into account several aspects of reality, such as the biotic, the psychic, the social, the economic, the legal, and the ethical aspect. But even then, this does not cover full reality. To illustrate this, we can use the example of map-making. Several different maps can be made of one and the same city: a topographical map, a tourist map, a water map, a military map, to mention some options. Each of these maps informs about the city, each in its own way. But even when I have studied all of these maps, I still do not know the full reality of this city. For that, I have to go into that city, walk around, touch, smell, and get an integrated impression of the reality of that city, in addition to the fragmented impressions coming from each of the maps. Science is the map-making activity that is very useful to get fragmented information about reality. But in technology we are dealing with reality in an integrated way. We cannot isolate any aspect or leave out any aspect, as reality is inherently complex and has all these aspects at the same time and in an integrated way.

This notion of reality as a complex whole of integrated aspects and science as an abstraction is very important for product innovation. Scientific research is a useful resource, but we also need integrated knowledge of reality. This knowledge comes from experience with reality, not from scientific research with its character of abstraction. This has important consequences for the knowledge portfolio of an industrial corporation. Both scientific knowledge, ranging from knowledge that is more phenomenon-oriented ('fundamental') to knowledge that is more practice-oriented (but still of an abstract nature), are needed in this portfolio. The role of each of these knowledge types varies between different product types. In the next section, three different types of relations between scientific knowledge and integrated knowledge of reality will be presented. For each product management will have to decide what type of product is at stake and organize the product development process according to the nature and needs of that particular type.

6.2 Experience-based technologies, interactive technologies, and applied science technologies

Philosopher of technology Andries Sarlemijn has developed a typology for product innovation processes in relation to scientific research. He distinguished three types of such processes and called them experience-based technologies, macrotechnologies, and microtechnologies (Sarlemijn and De Vries 1992). The latter two types are named after the spatial level at which the scientific knowledge refers to. The first type is named after the initial knowledge that guided the innovation. Thus, two different characteristics were used for labeling the technologies, which is not attractive. Hence, here these types will be renamed into experience-based technologies, interactive technologies, and applied science technologies. Each of these three types calls for different product management. We will now see how each of them can be characterized.

The first type of innovation is characterized by the scientific research serving as a sort of 'hindsight wisdom' rather than the engine of innovation. First, products are developed, based on experience with previous products, materials, and processes, without a proper understanding of the natural phenomena that determine the functioning on the product. Even with that new knowledge available, the product innovation continues to be driven by experience rather than by scientific research. A good example of that was the development of the self-puller corkscrew. This device was developed by, among other companies, the Dutch Brabantia household device company (De Vries1994). Because of a lawsuit, a lot of material is available about the development of that device. There are several physical forces that work within and on the corkscrew. But stories by the designers reveal that no systematic analyses of these forces using Newtonian mechanics were used in the development and later optimization of the design. Such analyses did exist and the designers were aware of an article about this in the *Scientific American*, but having read it, they smiled and said that if these analyses were to base the further design on such analyses, probably another corkscrew would never leave the factory. It would be a detour for them to use such analyses and it would not add anything essential to the experience on which they had based the design so far.

The second type is characterized by a back-and-forth movement between product development and scientific investigation, usually starting with product development. Once the product has been developed, the quest for improvement leads to scientific research with the idea that a better understanding of the natural phenomena will enable optimization of the product. A classic example is the steam engine that was developed without a proper understanding of what happens inside the machine but nevertheless functioned to a certain extent. The quest for improvement (higher power, higher efficiency) led to scientific research and resulted in thermodynamics being developed as a knowledge domain. This, in turn, led to improved steam engines. Further research led to improved thermodynamics, which led to even better steam engines, and etcetera. A similar example is geometrical optics that emerged from the need to improve lenses. Lenses had been developed for glasses (for reading), without a proper understanding of light fracture. But when these lenses were used for telescopes and microscopes, improvements were needed because of optical aberrations that made the lenses unusable in these new devices. Then research into light refraction at the transition between two media began and led to geometrical optics. This new knowledge was used to create better lenses and then a back-and-forth process similar to the steam engine example began.

The third type is again different. Here, the scientific research is a necessary precondition for product development and without it no real progress can be made. A good example of this is the development of the transistor at Bell Labs in the USA. Originally, the researchers tried to make a solid-state copy of the triode tube amplifier. At a certain moment it worked but not as predicted. Efforts to miniaturize the design failed. Later, the researchers turned to solid-state physics to understand what was going on in the device. Soon, they discovered the cause of the strange behavior of their artifact and found out that what they had designed was not a layer transistor but a field effect transistor (of course these names were added afterward). Thus, they suddenly ended up with two types of transistor that both worked. Understanding the field effect in their first design enabled them to optimize that type of transistor and the same knowledge also enabled them to design a proper layer transistor.

What does all this mean for product innovation management? It would be too simple to answer as follows: arrange a proper type of organization for each type of innovation. If it would be clear from the beginning to which type a new device belongs, that would be the way to go. But unfortunately, this is not always the case. Also it would be too simple to claim that innovations of products that already exist for a long time will behave like the first or the second type, because even an extremely novel type of technology such as nanotechnology seems to behave like the second type rather than the third. Nanoscientists put together a nano-device which then shows unexpected behavior for which research is needed to understand it and then improve the nano-device. A back-and-forth movement probably describes the process better than scientific research being a necessary precondition for developing the nano-devices (although the knowledge that is used to build the nano-devices of course is more than just experience). In fact, there is even an example of nanotechnology that is of the first type, namely stained glass. Makers of stained glass in the Middle Ages had no clue about gold atoms being inserted into the glass by their actions and it is only now that we understand the effect. Even now that knowledge does not mean much for stained glass–making, which is typical for the first type of product innovation.

Maybe the transistor example shows how to deal with the uncertainty about what type of innovation is at stake. The researchers began by using experience, but once they noticed that this did not lead to a desired result, they did not wait long to shift to another approach. Consequently, product innovation cannot be a process in which the whole sequence of development steps is predefined in advance. It is more like a step-by-step process in which constantly one evaluates whether or not the current approach (based on a belief about the type of innovation the product belongs to) is fruitful or not. This is what Sarlemijn called a 'piecemeal' approach to product innovation. The example of the Plumbicon television pickup tube at Philips Electronics shows that in this piecemeal process not only scientific research and design experience are involved but also knowledge about markets, legislation, and other social aspects of the device (Sarlemijn and De Vries 1992). This was already discussed in Chapter 3, so this will not be discussed here.

Coming back to the history of Philips Research, we can see a moving back and forth between different approaches. In the pre-WWII period, research in the laboratory and production in the factory were very much interwoven. The innovation management was clearly based on the idea that innovations behave like the second type in Sarlemijn's typology, and that approach appeared to be very fruitful for that period. After WWII, the company shifted toward an approach based on the third type of innovation. The corporate laboratory was seen as a great source of revolutionary innovations based on fundamental knowledge. This approach led to two great successes: the Plumbicon and LOCal Oxidation of Silicon (LOCOS), a procedure for making integrated circuits. Numerous other innovations, however, were frustrated because they did not fit in this scheme. Therefore, it was abandoned in the early 1970s, when the lead for product innovation was put into the hands of the product divisions and the laboratory became the deliverer of knowledge on demand. But the laboratory kept a limited amount of resources for fundamental research that still today contribute to product innovation.

6.3 Qualitative and quantitative research approaches

Apart from the variation in science-innovation relations as described in the previous section, there is also a variation in product innovation design methods. Often quantitative methods are used for this, both in the engineering and in the marketing realms. The advantage of quantitative methods is that well-known statistical methods are available to process the data. Besides that, numbers align well with the idea of control. Numbers suggest that the process is like a machine into which numbers can be fed to make the machine function as desired. The appreciation of quantitative data is, however, dependent on one's worldview. For some worldviews, quantitative data are all that matter. For other worldviews, quantitative data are useful but they are a reduction compared to full reality. A number of times already in this book, a non-reductionist approach has been suggested as more fruitful than a reductionist one. That is the case here, too. Qualitative data contain more than quantitative data. Particularly in situations where not much is yet known about reality, a qualitative approach can precede a more quantitative one. Qualitative data are useful for explorative research. A rich and deep description of phenomena can be obtained by qualitative research. The other side of the coin is that qualitative data are quite labor-intensive when it comes to analyzing them, and that is why, qualitative studies are almost always small-scale studies that do not allow for much generalization. For that, a follow-up of a more quantitative nature is needed. To give a simple and straightforward comparison of the possibilities of the two types of research, qualitative research can be used to prove the existence of a phenomenon. For instance, it can be proven qualitatively that customers exist that have certain desires regarding a new product. By using qualitative research, using interviews and focus groups, to mention just a few methods, rich and deep descriptions of these desires can be obtained, more than in a quantitative survey. But qualitative research cannot tell what percentage of customers have these desires. For that a more classic market survey is needed. There is a certain temptation to read qualitative data in a quantitative way. Although a small sample was used, researchers tend still to report in percentages of customers having certain opinions. Although this may be useful as a first impression, they certainly do not hold water when it comes to sound conclusions about the population and it is not desirable to build product innovation strategies on such conclusions. Even though it requires more effort to have a sequence of qualitative and quantitative studies, in the end it does pay off to invest in this.

The advantage of using these two types of research in innovation management is not to be taken for granted. This can be illustrated by the fact that in the *Oxford Handbook of Innovation Management* (Dodson et al, 2014), quantitative research is an entry in the index, whereas qualitative research is not. If in such a cornerstone publication the role of qualitative research is ill-recognized, this may be indicative for innovation management practice also. Of course, industrial corporations use customer focus groups in market research. But as stated above, there is a tendency to read qualitative data in a quantitative manner. Qualitative data are fundamentally different from quantitative data and the two should not be confused.

References

Dodson, M., Gann, D.M., and Phillips, N. (Eds.) (2014). *The Oxford Handbook of Innovation Management*. Oxford, UK: Oxford University Press.

Sarlemijn, A., and Vries, M.J. de (1992). The Piecemeal Rationality of Application-Oriented Research. An Analysis of the R&D-History Leading to the Invention of the Philips Plumbicon in the Philips Research Laboratories. In Kroes, P.A. and Bakker, M. (Eds.). *Technological Development and Science in the Industrial Age*. Dordrecht, The Netherlands: Kluwer Academic Publishers.

Vries, M.J. de (1994). Design Process Dynamics in an Experience-based Context: A Design Methodological Analysis of the Brabantia Corkscrew Development. *Technovation*, 14(7), 437–448.

Vries, M.J. de (2005). *80 Years of Research at the Philips Natuurkundig Laboratorium 1914–1994*. Amsterdam, The Netherlands: Pallas Publications (Amsterdam University Press).

7 Assessing the value, ethical implications, and impact of innovation

This chapter will discuss research that aims at assessing the value and possible impact of innovations. Particularly the way ethical values can be assessed will be discussed. As in Chapter 1, a reductionist approach will be avoided. Reducing ethics to consequences seems to be attractive, as consequences are mostly (at least partially) researchable. But that approach overlooks values that cannot be captured in consequences, and for those values, virtue and duty ethics have to come into the picture. Technology assessment as a widely used research approach for investigating possible consequences of innovations for the sake of innovation management will be discussed in this light.

7.1 Assessment as a responsibility in product innovation management

In this chapter, the assessment of proposed product designs will be discussed. This will be done in the broader context of the assessment of technologies for the sake of policy making. Before deciding about the various options for a new product, an assessment is desirable of the possible impact that the design will have. The scope of this assessment can be limited to the expected market position of the new product, but that is a rather narrow approach. The product will do much more than just be sold and bought. It may have an impact on social relations, national economies, globalization, the natural environment, and other aspects of life. Ideally a broad assessment is set up, of which the outcomes can still affect the product innovation management. It may lead to sophisticated choices between alternative designs or production processes, changes in the design, and certain market strategies. In a certain way, this can be seen as a social responsibility of the company. For this, the term Socially Responsible Innovation (SRI) has been introduced. The abbreviation SRI is also used for Socially

Responsible Investment, but that is much narrower than the topic of this chapter. SRI requires a multidimensional approach to innovation management in which scientific-technological and socio-economic issues are included (Flipse et al. 2014), and also, it cannot be a matter of industrial companies only but also entails political involvement (Rip, 2018). SRI brings us in the realm of ethics. In this chapter, an ethics that goes beyond risks and, in general, consequences will be proposed. Although a focus on consequences is attractive because consequences can be predicted through research to some extent, there are other relevant ethical concerns that need to be raised, related to more fundamental questions about what is morally permissible and what is to be rejected morally.

7.2 Technology assessment: forms and limitations

Although the term 'technology assessment' originates from a governmental context, this type of study is also relevant for product innovation management. It entails a systematic study into possible impacts of a new technological development. It usually has a broad scope and encompasses all sorts of impacts: on employment, economy, social relations, and the natural environment. It uses methods such as trend analysis and impact trees – methods that aim at investigating the future, based on the past and the present. Just like any other method, it has assumptions that need to be fulfilled in order for the method to yield reliable outcomes (see also Chapter 5). In this section, a distinction will be discussed that is rarely taken into account in current literature, yet is very important for a proper appreciation of what technology assessment can and cannot do. To explain it, we turn again to the aspects of reality that were introduced in Chapter 2, Section 2.2). In these aspects, two types of regularities (or 'laws') can be distinguished: 'natural' regularities and intentional regularities. The natural regularities are found in the 'lower' aspects (the numerical, spatial, kinematic, physical, and biotic). These regularities are studied in disciplines such as mathematics, physics, chemistry, and biology. These regularities do not depend on human intentionality. The actual number of parts a machine has does not depend on who counts them and with what purpose. The same holds for the space the machine occupies, the force it exerts on the factory floor, and the chemical properties of the materials it is made of. Similarly, the natural regularities of living organisms (such as cell behavior) also do not depend on the intentionality of the living being. I cannot influence the functioning of my cells. I can do things that make my heartbeat go up, but I cannot influence the mechanism through which that happens. These regularities have a high level of consistency and therefore allow for relatively accurate predictions. When I drop a screwdriver from a certain height, the laws of physics allow me to give a fairly accurate estimation of how long it will take for the screwdriver to hit the ground and with what force. In technology assessment, these regularities form part of the basis for trend analyses and impact trees. When we know current parameters and the change we have put into motion, these natural 'laws' allow us to predict the future, assuming that there will be no unexpected new parameters emerging. We happily use the fact that natural phenomena 'behave' in a strictly orderly way.

This is less the case with regularities that are based on phenomena in which human intentionality plays a role. Intentionality is a philosophical term that indicates not just intentions (purposes) but also attitudes, knowledge, emotions, efficacy, and the like. This sort of phenomena can be found in the 'higher' aspects. The 'laws' of psychology are not as absolute as those of physics, as people's behavior is not as regular as nature's behavior is. In psychology, models for relations between attitudes and behavior are developed, for instance. But such models cannot claim that the relations are absolute. Sometimes people let attitudes determine their behavior to a large extent, but sometimes people behave inconsistent and do things that go against their attitudes. I can have a very negative attitude toward smoking and yet be a stubborn smoker. This means that psychologists can do research that leads to expectations about customer behavior and later it can appear that customers do not behave according to the 'laws' that were found in psychology. The same holds for economy. That is why it is so difficult to predict the behavior of stock exchanges. Sometimes people panic and behave irrationally and do not act as could be expected from economic theories and models. Yet, technology assessment also depends on these kinds of regularities. It is important that the predictive value of different parts of a technology assessment is carefully evaluated by analyzing the type of regularities that were used. Predictions based on natural regularities can ascribe a high level of reliability, but predictions based on intentional regularities should be seen as much less accurate and certain.

The predictive value of technology assessment studies should also be evaluated against the knowledge of the factors that may have an impact on developments. This knowledge is never complete, but particularly in the early stages of a development it is difficult to establish which factors need to be taken into account (and about which regularities are known already or yet to be found). Unfortunately, this is also the period in which the decisions made, based on the outcomes of a technology assessment study, have the largest impact. As time goes on, it becomes more and more evident which factors are at stake. But also, the development goes on and the impact of decisions on the development becomes increasingly marginal. In technology assessment literature, this is called the Collingridge dilemma: either one makes decisions early on and they will have a great impact but are based on limited knowledge, or one waits until later when more knowledge is available but decisions have less impact as the technology has already taken shape and position. It is clear that the value of Collingridge's theory is limited in that it does not give clues about how to deal with this dilemma and therefore more is needed to deal wisely with this dilemma (Genus and Stirling, 2018). In Section 7.4, the need to complement the focus on impacts and consequences with other approaches in ethics will be spelled out as an answer to that need.

7.3 Environmental impact assessment: the issue of sustainability

One particular type of technology assessment is the assessment of impact on the sustainability of products and processes. So far in this book, not much has been written about sustainability. That certainly was not to deny the importance of this topic. Rather, this chapter on value-based assessment of products and technologies seemed to be an appropriate place for it. In the beginning of this chapter, it was already noticed that assessment is not something to be done only at the very end of the product development process. It should be a continuous activity throughout that process that guides it and helps to focus on certain values and avoid undesirable outcomes. That also holds for sustainability as a value that increasingly is seen as relevant by industrial companies, not in the least because unsustainable products and processes usually waste resources and thus money. Sustainability pays off, if not in the short run, then in the long run. Careful assessment of product and process properties that can be related to sustainability therefore is in the interest not only of society but also for the company.

Although the term sustainability is broader than the impact on the natural environment, in the first part of this section we will focus on that, as most methods are related to that. We are then in the area of Environmental (Impact) Assessment, a domain that is full of methods that need methodological consideration in order to be applied successfully. In Chapter 5, the assumptions in methods were pointed out as important for successful use of those methods. That holds here, too.

An example of that are Life Cycle Analyses (LCAs). These are methods for calculating the environmental impact of the whole life cycle of a product, starting from production to disposal or recycling. In the first place, LCAs assume that data are available about the material and energy use, and emissions to air, water, and soil for every phase in the life cycle. But also the method assumes that it is known where to start and where to end. That is by no means obvious. One can start with the production in the factory, but then the phase of raw material production is left out. Bringing in that phase creates new questions, such as should one also include the energy that was needed to make all the machines work that got the raw material out of nature? And if so, should one also include the production of the energy sources for those machines? And if so, should one also include the materials and energy needed for the plants that produced that energy, and etcetera. Clearly, there is no 'natural' or logical beginning. Neither is there a 'natural' ending, certainly not when the product is recycled, but also when it is left to decay with again a whole chain of causes and effects. Of course this matter can be solved by simply making an artificial 'cut' at the beginning and the end, but this needs to be carefully taken into account when comparing between alternatives for the product life cycle (and that is usually the purpose of LCAs).

The impact on the natural environment is only one of the impacts related to sustainability. The impact of the product on the sustainability of society is more than that. A product that has minimal impact on the natural environment but comes with a price that only few people can afford can hardly be called a sustainable product. Neither can that be claimed for a product that has little impact on the natural environment but is extremely difficult to use. Unsustainable technological developments can lead to national and international conflicts about an increasing lack of natural resources and an even more unbalanced distribution of resources over industrialized and developing countries. Those are impacts that go way beyond the damage done to the natural environment in itself. In a good Environmental (Impact) Assessment, the human and social aspects of sustainability are included (Morgan, 2012). Even this perspective of impacts is not all there is. A full assessment also needs to include the managerial aspects of how to set up a product innovation management that leads to sustainable products. This broader perspective on sustainability assessment has been elaborated in what is called Strategic Environmental Assessment (SEA) (Bram Noble, 2017; Schiederig et al, 2012). That type of assessment also includes the way one deals with the general public. After all, they need to be convinced that sustainable products are worth their price, or vice versa, it is the general public that in its purchasing behavior displays a sense for the value of sustainability that needs to be addressed by the industry. In Chapter 3, we have already discussed the way product innovation management involves dealing with a whole network of social actors. That most certainly also holds in the realm of sustainability. Governments, consumers, special interest organizations such as Greenpeace, actors in supply chains, they all play a role in the whole process. They all hold values: protecting public justice, protecting the well-being of nature and humans, making profit, and all of these need to be taken into account if a product should not 'stumble' somewhere in its life cycle. Most of these values are moral in nature. They are related to what we consider to be 'good' rather than 'evil' or 'bad'. So far, these values were assessed based on impacts and consequences. The ethical stance that focuses on that is often called 'consequentionalism'. This is, however, not the only way to assess moral values. It is a limited one in that it assumes that we can predict consequences, which is increasingly problematic the newer and more innovative the product is. In the next section, other approaches to assessing moral values are discussed.

7.4 Non-consequentionalist forms of ethics

In any act, three components can be distinguished: the person who acts, the act itself, and the consequences of the act. Each of these three can be the focus of a moral assessment. Consequentionalism focuses on the consequences of the act. If the consequences of the act are good (in most cases that will be a total good sum of good and bad sub-consequences), performing the act is morally permissible. A second approach focuses on the person who performs the act. If that person can be said to be 'a good person', then performing the act is morally permissible. This approach is usually called virtue ethics, whereby virtues are good characteristics of a person. A third approach focuses on the act itself. Some acts are intrinsically good or bad. Only good acts are morally permissible. It is everyone's duty to perform only such acts, irrespective of consequences. That approach is generally indicated as duty ethics. Most people will not opt for an either–or approach but combine all three perspectives and make an overall consideration that includes consequences, virtues, and duties. That combined approach can be called 'responsibility ethics', because it contains the word 'response' that refers to the account one can provide when asked for one's motive to perform or not to perform a certain act.

What does all that mean for product innovation management? Throughout this book, reductionist approaches to reality have been shown to have risks in that crucial aspects of reality can be overlooked, which may lead to problematic product development. In this final chapter, again a plea is made for a broad approach. Although consequences are attractive as a focus as they can be investigated scientifically, it is always worth considering that there may be acts that one would not like to perform, even if there are no obvious 'bad' consequences. Here again worldviews come in. A certain worldview may include a view on humans that makes certain acts with and on humans intrinsically problematic (such as tinkering with their DNA or treating them as mindless machines). The same holds for certain views on nature. Even if there is no real harm on humans when certain types of animals or plants die out, that loss of biodiversity can still be seen as morally bad. Technologies that treat humans as a 'bag of molecules' that can be manipulated in any way as long as no perceivable harm is done can be problematic in a worldview in which humans are more than material objects and have a human dignity that needs to be respected. New products and technologies can be assessed on whether or not they ignore human dignity, even when the human does not notice anything themselves. Finally, product innovation management can also involve virtues, such as our motivations to develop new products. Different people involved in product development will have different motives for what they do, and in a virtue ethics approach, there is room for different people with different motivations.

7.5 Final remarks

We have now completed our travel along the various phases and aspects of product innovation management for which the philosophy of technology has a message. Probably the reader is left with the impression that this does not make things simpler. Philosophy tends to ask more questions than provide answers. But asking the right questions can prevent taking answers for granted that in fact are simplifications that endanger the desirable outcome of product innovation management: a product or process that is a meaningful contribution to society, designed and produced by people who feel that there was room for their professional qualities and expertise. Rather than forcing everyone and everything in a mold that nicely simplifies but simultaneously feels like a straitjacket, this book suggests to see complexity as a wealth of possibilities. Then philosophy is more than "when he who hears does not know what he who speaks means, and when he who speaks does not know what he himself means", to quote the beginning of Chapter 1. Or in the words of the psychologist Kurt Lewin (perhaps), or physicist Albert Einstein (perhaps), or physicist Ludwig Bolzmann (perhaps) or any other person to whom this quote has already been contributed: "Nothing is more practical than a good theory".

References

Bram Noble, Kelechi Nwanekezie (2017). Conceptualizing Strategic Environmental Assessment: Principles, Approaches and Research Directions. *Environmental Impact Assessment Review*, 62, 165–173.

Flipse, Steven, M., van der Sanden, Maarten, C.A., Radstake, Maud, De Winde, Johannes H., and Osseweijer, Patricia. (2014). The DNA of Socially Responsible Innovation. Social and Natural Scientists Need to Establish Mutual Understanding and a Common Language to Efficiently Work Together. *EMBO Reports*, 15, 134–137.

Genus, A., and Stirling, A. (2018). Collingridge and the Dilemma of Control: Towards Responsible and Accountable Innovation. *Research Policy*, 47(1), 61–69.

Morgan, Richard K. (2012). Environmental Impact Assessment: The State of the Art. *Impact Assessment and Project Appraisal*, 30(1), 5–14.

Rip, A. (2018) Constructive Technology Assessment. In Rip, A (Ed.), *Futures of Science and Technology in Society. Technikzukünfte, Wissenschaft und Gesellschaft / Futures of Technology, Science and Society.* Wiesbaden, Germany: Springer VS.

Schiederig, T., Tietze, F., and Herstatt, C. (2012). Green Innovation in Technology and Innovation Management. *R&D Management*, 42, 180–192.

Index

Printed in the United States
by Baker & Taylor Publisher Services